Everything Added to Food in the United States

U.S. Food and Drug Administration

C. K. SMOLEY

Library of Congress Cataloging-in-Publication Data

Everything added to food in the United States / Center for Food Safety
and Applied Nutrition. Division of Toxicological Review and
Evaluation. U.S. Food and Drug Administration.
    p. cm.
  ISBN 0-8493-8723-X
  1. Food additives. I. Center for Food Safety and Applied
Nutrition (U.S.). Division of Toxicological Review and Evaluation.
TX553.A3E94   1993
664'.06--dc20   93-18072
                                                      CIP

Direct all inquiries to CRC Press, Inc.
2000 Corporate Blvd. N.W.
Boca Raton, Florida 33431

PRINTED IN THE UNITED STATES OF AMERICA
  3 4 5 6 7 8 9 0

Printed on acid-free paper

The information in this book is derived from the files of the FDA. It contains 2922 total food additives of which 1755 are regulated food additives, including direct, secondary direct, color and Generally Recognized As Safe (GRAS) additives. In addition, the book contains administrative and chemical information on 1167 additional such substances. Thus, the 2922 total additives together comprise EVERYTHING ADDED TO FOOD IN THE UNITED STATES.

The book is arranged alphabetically and contains selected information on the 2922 food additives. Definitions for the labels are:

MAINTERM: Name of the chemical as recognized by the Center for Food Safety and Applied Nutrition (CFSAN) of the U.S. Food and Drug Administration.

CAS: Chemical Abstract Service (CAS) Registry code for the chemical or CAS-like code assigned by CFSAN to those substances that do not have a CAS number (977nnnnnn-series)

REGNUM: Regulation numbers in Title 21 of the U.S. Code of Federal Regulations where the chemical is listed.

MAINTERM	CAS	REGNUM
ACACIA, GUM (ACACIA SENEGAL (L.) WILLD.)	009000015	184.1330 169.179 169.182 170.3(0)(14) 170.3(0)(28)
ACESULFAME POTASSIUM	055589623	172.800
ACETAL	000105577	172.515
ACETALDEHYDE	000075070	182.60
ACETALDEHYDE, BUTYL PHENETHYL ACETAL	064577919	
ACETALDEHYDE PHENETHYL PROPYL ACETAL	007493574	172.515
ACETANISOLE	000100061	172.515
ACETIC ACID	000064197	182.1 182.70 PART 133 172.814 184.1005 73.85 178.1010
ACETIC ANHYDRIDE	000108247	172.892
ACETOIN	000513860	182.60
ACETOLEIN	028060904	
ACETONE	000067641	175.105 27 CFR 212.30 40 CFR 180.1001 73.615 176.300 182.99 175.320 73.30 173.210 73.345 73.1
ACETONE PEROXIDES	001336170	137.105 172.802 137.155 137.165 137.160 137.170 137.175 137.180 137.185

MAINTERM	CAS	REGNUM
ACETOPHENONE	000098862	172.515
ACETOSTEARIN	027177851	
6-ACETOXYDIHYDROTHEASPIRANE	072541094	
4-(P-ACETOXYPHENYL)-2-BUTANONE	003572063	
4-ACETYL-6-TERT-BUTYL-1,1-DIMETHYL-INDANE	013171001	
3-ACETYL-2,5-DIMETHYLFURAN	010599709	
2-ACETYL-3,(5 OR 6)-DIMETHYLPYRAZINE, MIXTURE OF ISOMERS	977043632	
3-ACETYL-2,5-DIMETHYLTHIOPHENE	002530101	
2-ACETYL-3-ETHYLPYRAZINE	032974928	
N-ACETYL-L-METHIONINE	000065827	172.372
ACETYL METHYL CARBINYL ACETATE	004906245	
2-ACETYL-5-METHYLFURAN	001193799	
4-ACETYL-2-METHYLPYRIMIDINE	067860382	
ACETYLPYRAZINE	022047252	
2-ACETYLPYRIDINE	001122629	
3-ACETYLPYRIDINE	000350038	
2-ACETYLTHIAZOLE	024295032	
ACONITIC ACID	000499127	184.1007
ACROLEIN	000107028	172.892 176.300
ACRYLAMIDE	000079061	
ACRYLAMIDE-ACRYLIC ACID RESIN	009003069	173.5
ACRYLAMIDE-SODIUM ACRYLATE RESIN	025085023	173.310 172.710 173.5
ACRYLIC ACID	000079107	175.105 175.300

MAINTERM	CAS	REGNUM
		175.360
		175.365
		175.380
		175.390
		176.170
		177.1010
		177.1210
		177.1350
		177.2260
		178.3790
ACRYLIC ACID-2-ACRYLAMIDO-2-METHYL PROPANE SULFONIC ACID COPOLYMER	040623754	173.310
ACTIVATED CARBON	064365113	
ADIPIC ACID	000124049	184.1009 131.144
ADIPIC ANHYDRIDE	002035758	172.892
AGAR (GELIDIUM SPP.)	009002180	150.161 150.141 184.1115
B&-ALANINE	000107959	
L-ALANINE	000056417	172.320
DL-ALANINE	000302727	172.540
ALBUMIN	977005723	
ALCOHOL, DENATURED FORMULA 23A	977021649	73.1
ALCOHOL SDA-3A	977021592	73.1
ALFALFA, EXTRACT (MEDICAGO SATIVA L.)	977025209	182.20
ALFALFA, HERB AND SEED (MEDICAGO SATIVA L.)	977092935	182.10
ALGAE, BROWN, EXTRACT (MACROCYSTIS AND LAMINARIA SPP.)	977026928	184.1120 172.365
ALGAE, RED, EXTRACT (PORPHYRA SPP. AND GLOIOPELTIS FURCATA AND RHODYMENIA PALMATA (L.))	977090042	184.1121
ALGAE, RED (PORPHYRA SPP. AND GLOIOPELTIS FURCATA AND RHODYMENIA PALMATA (L.))	977007741	184.1121
ALGINATE, AMMONIUM	009005349	173.310 184.1133

MAINTERM	CAS	REGNUM
ALGINATE, CALCIUM	009005350	184.1187
ALGINATE, POTASSIUM	009005361	184.1610
ALGINATE, SODIUM	009005383	133.133 150.161 150.141 133.179 133.178 133.162 133.134 173.310 184.1724
ALGINATE, SODIUM CALCIUM	977054297	
ALGINIC ACID	009005327	184.1011
ALKANET ROOT, EXTRACT (ALKANNA TINCTORIA TAUSCH)	023444657	
ALKANOLAMIDE OF COCONUT OIL FATTY ACIDS AND DIETHANOLAMINE	068603429	173.322
N-ALKYL(C8-C18 FROM COCONUT OIL) AMINE ACETATE	977075630	172.710
ALKYLENE OXIDE ADDUCTS OF ALKYL ALCOHOLS/PHOSPHATE ESTERS OF SAME, MIXTURE	977043745	173.315
A&-ALKYL-W&-HYDROXY-POLY(OXYETHYLE-NE)	977044317	173.322
ALLSPICE, OIL (PIMENTA OFFICINALIS LINDL.)	008006777	182.20
ALLSPICE, OLEORESIN (PIMENTA OFFICINALIS LINDL.)	977017870	182.20
ALLSPICE (PIMENTA OFFICINALIS LINDL.)	977051721	182.10
ALLYL ANTHRANILATE	007493632	172.515
ALLYL BUTYRATE	002051787	172.515
ALLYL CINNAMATE	001866315	172.515
ALLYL CROTONATE	020474935	
ALLYL CYCLOHEXANEACETATE	004728829	172.515
ALLYL CYCLOHEXANEBUTYRATE	007493654	172.515
ALLYL CYCLOHEXANEHEXANOATE	007493665	172.515

MAINTERM	CAS	REGNUM
ALLYL CYCLOHEXANEPROPIONATE	002705875	172.515
ALLYL CYCLOHEXANEVALERATE	007493687	172.515
4-ALLYL-2,6-DIMETHOXYPHENOL	006627889	
ALLYL DISULFIDE	002179579	172.515
ALLYL 2-ETHYLBUTYRATE	007493698	172.515
ALLYL 2-FUROATE	004208495	
ALLYL HEPTANOATE	000142198	
ALLYL HEXANOATE	000123682	172.515
ALLYL HEXENOATE	977075301	
ALLYL A&-IONONE	000079787	172.515
ALLYL ISOTHIOCYANATE	000057067	172.515
ALLYL ISOVALERATE	002835394	172.515
ALLYL MERCAPTAN	000870235	172.515
ALLYL METHYL DISULFIDE	002179580	
ALLYL METHYL TRISULFIDE	034135858	
ALLYL NONANOATE	007493723	172.515
ALLYL OCTANOATE	004230971	172.515
ALLYL PHENOXYACETATE	007493745	172.515
ALLYL PHENYLACETATE	001797746	172.515
ALLYL PROPIONATE	002408200	172.515
ALLYL SORBATE	007493756	172.515
ALLYL SULFIDE	000592881	172.515
ALLYL THIOPROPIONATE	041820228	
ALLYL TIGLATE	007493712	172.515
ALLYL 10-UNDECENOATE	007493767	172.515

MAINTERM	CAS	REGNUM
ALMONDS, BITTER, OIL (FFPA) (PRUNUS SPP.)	008013761	182.20
ALOE, EXTRACT (ALOE SPP.)	084837081	172.510
ALTHEA FLOWERS (ALTHEA OFFICINALIS L.)	977052713	172.510
ALTHEA ROOT (ALTHEA OFFICINALIS L.)	977005756	172.510
ALUM (DOUBLE SULFATE OF AL AND NH4, K, OR NA)	977007616	182.90
ALUMINUM AMMONIUM SULFATE	007784261	182.1127 182.90
ALUMINUM CALCIUM SILICATE	001327395	182.2122 169.179 169.182
ALUMINUM CAPRATE	022620935	172.863
ALUMINUM CAPRYLATE	006028575	172.863
ALUMINUM HYDROXIDE	021645512	182.90
ALUMINUM LAURATE	007230935	172.863
ALUMINUM MYRISTATE	004040500	172.863
ALUMINUM NICOTINATE	001976289	172.310
ALUMINUM OLEATE	000688379	182.90 172.863
ALUMINUM PALMITATE	000555351	172.863 182.90
ALUMINUM POTASSIUM SULFATE	007784249	133.165 133.141 133.111 133.106 133.102 133.183 133.181 137.105 182.1129 182.90 133.195 137.155 137.165 137.160 137.170 137.175 137.180 137.185
ALUMINUM SALTS OF FATTY ACIDS	977089512	172.863

MAINTERM	CAS	REGNUM
ALUMINUM SODIUM SULFATE	007784283	182.90 182.1131
ALUMINUM STEARATE	000637127	172.863 173.340
ALUMINUM SULFATE	010043013	182.1125 172.892
AMBERGRIS, TINCTURE	977023087	182.50
AMBRETTE, ABSOLUTE, OIL (HIBISCUS ABELMOSCHUS L.)	977017790	182.20
AMBRETTE SEED (HIBISCUS ABELMOSCHUS L.)	977052202	182.10
AMBRETTE SEED, OIL (HIBISCUS ABELMOSCHUS L.)	008015621	182.20
AMBRETTE, TINCTURE (HIBISCUS ABELMOSCHUS L.)	977017789	182.20
P-AMINOBENZOIC ACID	000150130	
DL-(3-AMINO-3-CARBOXYPROPYL)DIMETH-YLSULFONIUM CHLORIDE	001115840	
AMINO TRI(METHYLENE PHOSPHONIC ACID), SODIUM SALT	020592852	
AMMONIUM ACETATE	000631618	
AMMONIUM BICARBONATE	001066337	163.110 184.1135
AMMONIUM CARBONATE	008000735	163.110 184.1137
AMMONIUM CASEINATE	009005429	135.110 135.140
AMMONIUM CHLORIDE	012125029	184.1138
AMMONIUM CITRATE	003012655	175.105 175.300 177.1350 175.380 175.390 181.29 184.1140
AMMONIUM GLUCONATE	002554043	
AMMONIUM HYDROXIDE	001336216	184.1139 163.110

MAINTERM	CAS	REGNUM
AMMONIUM ISOVALERATE	007563339	172.515
AMMONIUM PERSULFATE	007727540	172.892
AMMONIUM PHOSPHATE, DIBASIC	007783280	573.320 184.1141B
AMMONIUM PHOSPHATE, MONOBASIC	007722761	184.1141A
AMMONIUM SULFATE	007783202	184.1143
AMMONIUM SULFIDE	012135761	172.515
AMYL ALCOHOL	000071410	172.515
AMYLASE FROM ASPERGILLUS FLAVUS	977032055	
AMYLASE FROM ASPERGILLUS NIGER	888284400	
AMYLASE FROM ASPERGILLUS ORYZAE	977082953	137.105 137.200 137.205 137.155 137.165 137.160 137.170 137.175 137.180 137.185
AMYLASE FROM BACILLUS SUBTILIS	977028311	
AMYL BUTYRATE	000540181	172.515
A&-AMYLCINNAMALDEHYDE	000122407	172.515
A&-AMYLCINNAMALDEHYDE DIMETHYL ACETAL	000091872	172.515
A&-AMYLCINNAMYL ACETATE	007493789	172.515
A&-AMYLCINNAMYL ALCOHOL	000101859	172.515
A&-AMYLCINNAMYL FORMATE	007493790	172.515
A&-AMYLCINNAMYL ISOVALERATE	007493803	172.515
AMYL DECANOATE	005933879	
AMYL FORMATE	000638493	172.515
AMYL 2-FUROATE	004996489	
AMYL HEPTANOATE	007493825	172.515

MAINTERM	CAS	REGNUM
AMYL HEXANOATE	000540078	172.515
2-AMYL-5 OR 6-KETO-1,4-DIOXANE	065504963	
AMYL OCTANOATE	000638255	172.515
AMYLOGLUCOSIDASE FROM RHIZOPUS NIVEUS	977080402	173.110
AMYL SALICYLATE	002050080	
AMYRIS (AMYRIS BALSAMIFERA L.)	977059690	172.510
AMYRIS, OIL (AMYRIS BALSAMIFERA L.)	008015654	172.510
TRANS-ANETHOLE	004180238	182.60
ANGELICA ROOT (ANGELICA SPP.)	977050068	182.10
ANGELICA ROOT, EXTRACT (ANGELICA ARCHANGELICA L.)	977032497	182.20
ANGELICA ROOT, OIL (ANGELICA ARCHANGELICA L.)	008015643	182.20
ANGELICA SEED (ANGELICA SPP.)	977050079	182.10
ANGELICA SEED, EXTRACT (ANGELICA ARCHANGELICA L.)	977032500	182.20
ANGELICA SEED, OIL (ANGELICA ARCHANGELICA L.)	977050080	182.20
ANGELICA STEM, OIL (ANGELICA ARCHANGELICA L.)	977032486	182.20
ANGOLA WEED (ROCCELLA FUCIFORMIS ACH.)	977038440	172.510
ANGOSTURA, EXTRACT (GALIPEA OFFINCINALIS HANCOCK)	068916121	182.20
ANGOSTURA (GALIPEA OFFINCINALIS HANCOCK)	977000228	182.10
ANISE, OIL (PIMPINELLA ANISUM L.)	008007703	182.20
ANISE (PIMPINELLA ANISUM L.)	977007650	182.10
ANISE, STAR (ILLICIUM VERUM HOOK, F.)	977052166	182.10
ANISE, STAR, OIL (ILLICIUM VERUM HOOK, F.)	068952432	

MAINTERM	CAS	REGNUM
ANISIC ACID	001335086	
ANISOLE	000100663	172.515
ANISYL ACETATE	000104212	172.515
ANISYL ALCOHOL	000105135	172.515
ANISYL BUTYRATE	006963560	172.515
ANISYL FORMATE	000122918	172.515
ANISYL PHENYLACETATE	000102170	172.515
ANISYL PROPIONATE	007549339	172.515
ANNATTO, EXTRACT (BIXA ORELLANA L.)	008015676	73.30 73.2030 73.1030
ANNATTO, SEED (BIXA ORELLANA L.)	001393631	
ANTHRACITE COAL, SULFONATED	069013203	173.25
B&-APO-8'-CAROTENAL	001107262	73.90
APPLE ESSENCE, NATURAL	977090735	
APRICOT KERNEL, OIL (PRUNUS ARMENIACA L.)	072869693	182.40
A&-(P-(1,1,3,3-TETRAMETHYLBUTYL)PH-ENYL)-W&-HYDROXYPOLY(OXYETHYLENE)(-GREATER THAN 1 MOL)	009002931	172.710
ARABINOGALACTAN	009036662	172.610 172.230
L-ARABINOSE	005328370	
L-ARGININE	000074793	172.320
ARNICA FLOWERS (ARNICA SPP.)	977000273	172.510
ARTEMISIA (ARTEMISIA SPP.)	977052735	172.510
ARTEMISIA EXTRACT	977032373	172.510
ARTEMISIA OIL	008008933	172.510
ARTICHOKE LEAVES (CYNARA SCOLYMUS L.)	977020475	172.510

MAINTERM	CAS	REGNUM
ASAFETIDA, FLUID EXTRACT (FERULA ASSAFOETIDA L.)	977038462	182.20
ASAFETIDA, GUM (FERULA ASSAFOETIDA L.)	009000048	
ASAFETIDA, OIL (FERULA ASSAFOETIDA L.)	072869706	182.20
ASCORBIC ACID	000050817	182.3013 146.113 145.170 145.110 137.200 145.115 182.5013 145.135 161.175 156.145 155.200 137.105 150.161 150.141 146.187 145.116 145.171 145.136 137.205 137.155 182.8013 137.165 137.160 137.170 137.175 137.180 137.185 240.1044
ASCORBYL PALMITATE	000137666	166.110 182.3149
ASCORBYL STEARATE	025395668	166.110
L-ASPARAGINE	000070473	172.320
ASPARAGUS, SEED AND ROOT, EXTRACT	977082964	
ASPARTAME	022839470	172.804
L-ASPARTIC ACID	000056848	172.320
ASPERGILLUS NIGER FOR FERMENTATION PRODUCTION OF CITRIC ACID	009001234	173.280
AZODICARBONAMIDE	000123773	172.806 137.105 136.110 137.200 137.205 136.115 136.130 136.160 136.165

MAINTERM	CAS	REGNUM
		136.180
		137.155
		137.165
		137.160
		137.170
		137.175
		137.180
		137.185
BACTERIAL CATALASE FROM MICROCOCCUS LYSODEIKTICUS	977050353	173.135
BAKER'S YEAST PROTEIN	977014133	172.325
BAKER'S YEAST GLYCAN	977014122	172.898
BALM LEAVES, EXTRACT (MELISSA OFFICINALIS L.)	977060186	182.20
BALM LEAVES (MELISSA OFFICINALIS L.)	977090746	182.10
BALM (MELISSA OFFICINALIS L.)	977051083	182.10
BALM, OIL (MELISSA OFFICINALIS L.)	008014719	182.20
BALSAM FIR, OIL (ABIES BALSAMEA (L.) MILL.)	008024155	
BALSAM FIR, OLEORESIN (ABIES BALSAMEA (L.) MILL.)	977017814	
BALSAM, PERU (MYROXYLON PEREIRAE KLOTZSCH)	008007009	182.20
BALSAM, PERU, OIL (MYROXYLON PEREIRAE KLOTZSCH)	977136927	182.20
BASIL BUSH (OCIMUM MINIMUM L.)	977051550	182.10
BASIL (OCIMUM BASILICUM L.)	977050148	182.10
BASIL, OIL (OCIMUM BASILICUM L.)	008015734	182.20
BASIL, OLEORESIN (OCIMUM BASILICUM L.)	977017825	182.20
BAY (LAURUS NOBILIS L.)	977050159	182.10
BAY LEAVES, SWEET, EXTRACT (LAURUS NOBILIS L.)	977090791	182.20
BAY LEAVES, SWEET, OIL (LAURUS NOBILIS L.)	008007485	182.20
BAY LEAVES, WEST INDIAN, EXTRACT (PIMENTA ACRIS KOSTEL)	977090779	NOT REGULATED

MAINTERM	CAS	REGNUM
BAY LEAVES, WEST INDIAN, OIL (PIMENTA RACEMOSA (MILL.) J.W. MOORE)	008006788	182.20
BAY LEAVES, WEST INDIAN, OLEORESIN (PIMENTA ACRIS KOSTEL)	977090780	NOT REGULATED
BEECHWOOD, CREOSOTE (FAGUS SPP.)	008021394	172.515
BEESWAX	008012893	184.1973
BEESWAX, BLEACHED	008006404	184.1973
BENTONITE	001302789	184.1155
BENZALDEHYDE	000100527	182.60
BENZALDEHYDE DIMETHYL ACETAL	001125888	172.515
BENZALDEHYDE GLYCERYL ACETAL	001319886	172.515
BENZALDEHYDE PROPYLENE GLYCOL ACETAL	002568254	172.515
BENZENE	000071432	172.560 175.105
BENZENETHIOL	000108985	172.515
2-BENZOFURANCARBOXALDEHYDE	004265161	
BENZOIC ACID	000065850	150.161 150.141 184.1021 166.110 166.40
BENZOIN	000119539	172.515
BENZOIN, RESIN (STYRAX SPP.)	009000059	73.1 172.510
BENZOPHENONE	000119619	172.515
BENZOTHIAZOLE	000095169	
BENZOYL PEROXIDE	000094360	133.195 133.183 133.181 133.111 133.106 133.102 137.105 133.165 133.141 137.155

MAINTERM	CAS	REGNUM
		172.814
		137.165
		137.160
		137.170
		137.175
		137.180
		137.185
		184.1157
BENZYL ACETATE	000140114	172.515
BENZYL ACETOACETATE	005396894	172.515
BENZYL ALCOHOL	000100516	172.515
		175.105
		175.300
		177.1210
BENZYL BENZOATE	000120514	172.515
BENZYL BUTYL ETHER	000588670	172.515
BENZYL BUTYRATE	000103377	172.515
BENZYL CINNAMATE	000103413	172.515
BENZYL 2,3-DIMETHYLCROTONATE	007492695	172.515
BENZYL DISULFIDE	000150607	172.515
BENZYL ETHYL ETHER	000539300	172.515
BENZYL FORMATE	000104574	172.515
3-BENZYL-4-HEPTANONE	007492377	172.515
BENZYL ISOBUTYRATE	000103286	172.515
BENZYL ISOVALERATE	000103388	172.515
BENZYL MERCAPTAN	000100538	172.515
BENZYL METHOXYETHYL ACETAL	007492399	172.515
BENZYL TRANS-2-METHYL-2-BUTENOATE	037526888	
BENZYL METHYL SULFIDE	000766927	
BENZYL PHENYLACETATE	000102169	172.515
BENZYL PROPIONATE	000122634	172.515
BENZYL SALICYLATE	000118581	172.515

MAINTERM	CAS	REGNUM
BERGAMOT, OIL (CITRUS AURANTIUM L. SUBSP. BERGAMIA WRIGHT ET ARN.)	008007758	182.20
BIOTIN	000058855	182.5159 182.8159
BIPHENYL	000092524	
BIRCH, SWEET, OIL (BETULA LENTA L.)	068917500	
BIRCH TAR, OIL (BETULA PENDULA ROTH AND RELATED BETULA SPP.)	008001885	172.515
BISABOLENE	000495625	
BIS(2,5-DIMETHYL-3-FURYL) DISULFIDE	028588730	
BIS(2-METHYL-3-FURYL) DISULFIDE	028588752	
BIS(2-METHYL-3-FURYL) TETRASULFIDE	028588763	
BLACKBERRY BARK, EXTRACT (RUBUS, SPP. OF SECTION EUBATUS)	977047532	172.510
BLACKBERRY FRUIT EXTRACT	977025232	
BOIS DE ROSE, OIL (ANIBA ROSAEODORA DUCKE)	008015778	182.20
BOLDUS LEAVES (PEUMUS BOLDUS MOL.)	977052757	172.510
BONITO, DRIED	977138718	
BORAX	001303964	
BORIC ACID	010043353	175.105 176.180
BORNEOL	000507700	172.515
BORNYL ACETATE	000076493	172.515
BORNYL FORMATE	007492413	172.515
BORNYL ISOVALERATE	000076506	172.515
BORNYL VALERATE	007549419	172.515
BORONIA, ABSOLUTE (BORONIA MEGASTIGMA NEES)	008053336	172.510
BOUILLON, VEGETABLE, SMOKE	977090815	

MAINTERM	CAS	REGNUM
B&-BOURBONENE	005208593	172.515
BROMELAIN	009001007	
BROMINATED VEGETABLE OIL	008016942	180.3
BRYONIA ROOT (BRYONIA SPP.)	977000499	172.510
BUCHU LEAVES EXTRACT	977009827	172.510
BUCHU LEAVES, OIL (BAROSMA SPP.)	068650464	172.510
BUCKBEAN LEAVES, EXTRACT (MENYANTHES TRIFOLIATA L.)	977038520	172.510
BUCKBEAN LEAVES (MENYANTHES TRIFOLIATA L.)	977038519	172.510
BUTADIENE-STYRENE RUBBER	009003558	172.615
BUTANE	000106978	173.350 184.1165
1,2-BUTANEDITHIOL	016128680	
1,3-BUTANEDITHIOL	024330527	
2,3-BUTANEDITHIOL	004532643	
1-BUTANETHIOL	000109795	
2-BUTANOL	000078922	172.515
2-BUTANONE	000078933	172.515
BUTAN-3-ONE-2-YL BUTANOATE	084642615	
BUTTER ACIDS	085536250	172.515
BUTTER ESTERS	977019263	172.515
BUTTER FAT, ENZYME-MODIFIED, WITH ADDED BUTYRIC ACID	977093256	
BUTTER STARTER DISTILLATE	977019274	184.1848
BUTYL ACETATE	000123864	172.515
BUTYL ACETOACETATE	000591606	172.515
BUTYL ALCOHOL	000071363	172.515 172.560 73.1

MAINTERM	CAS	REGNUM
BUTYLAMINE	000109739	
BUTYL ANTHRANILATE	007756969	172.515
BUTYLATED HYDROXYANISOLE	025013165	182.3169
		166.110
		172.515
		172.615
		173.340
		172.110
		181.24
		175.105
		175.125
		175.300
		175.380
		175.390
		176.170
		176.210
		177.1010
		177.1210
		177.1350
		178.3120
		178.3570
		179.45
BUTYLATED HYDROXYTOLUENE	000128370	182.3173
		173.340
		172.115
		137.350
		172.615
		166.110
		181.24
		175.105
		175.125
		175.300
		175.380
		175.390
		176.170
		176.210
		177.1010
		177.1210
		177.1350
		177.2600
		178.3120
		178.3570
		179.45
2-BUTYL-2-BUTENAL	025409089	
BUTYL BUTYRATE	000109217	172.515
BUTYL BUTYRYLLACTATE	007492708	172.515
A&-BUTYLCINNAMALDEHYDE	007492446	172.515
BUTYL CINNAMATE	000538658	172.515
2-SEC-BUTYLCYCLOHEXANONE	014765301	
BUTYL 2-DECENOATE	007492457	172.515
2-(2-BUTYL)-4,5-DIMETHYL-3-THIAZOL-INE	065894828	

MAINTERM	CAS	REGNUM
1,3-BUTYLENE GLYCOL	000107880	173.220
SEC-BUTYL ETHYL ETHER	002679870	
BUTYL ETHYL MALONATE	017373841	172.515
BUTYL FORMATE	000592847	172.515
BUTYL HEPTANOATE	005454284	172.515
BUTYL HEXANOATE	000626824	172.515
TERT-BUTYLHYDROQUINONE	001948330	172.185 177.2420
BUTYL P-HYDROXYBENZOATE	000094268	172.515
3-BUTYLIDENEPHTHALIDE	000551086	
BUTYL ISOBUTYRATE	000097870	172.515
BUTYL ISOVALERATE	000109193	172.515
2-BUTYL-5 OR 6-KETO-1,4-DIOXANE	065504952	
BUTYL LACTATE	000138227	172.515
BUTYL LAURATE	000106183	172.515
BUTYL LEVULINATE	002052155	172.515
N-BUTYL 2-METHYLBUTYRATE	015706737	
BUTYL PHENYLACETATE	000122430	172.515
3-N-BUTYLPHTHALIDE	006066495	
BUTYL PROPIONATE	000590012	172.515
BUTYL SALICYLATE	002052144	
BUTYL STEARATE	000123955	172.515 173.340
BUTYL SULFIDE	000544401	172.515
BUTYL 10-UNDECENOATE	000109422	172.515
BUTYL VALERATE	000591684	172.515

MAINTERM	CAS	REGNUM
BUTYRALDEHYDE	000123728	172.515
BUTYRIC ACID	000107926	182.60
CADINENE	029350730	172.515
CAFFEINE	000058082	182.1180 165.175
CAJEPUT, OIL (MELALEUCA LEUCADENDRON L.)	008008988	172.510
CALAMUS--PROHIBITED	977022904	189.110
CALAMUS OIL--PROHIBITED	008015790	189.110
CALCIUM ACETATE	000062544	182.6197 184.1185
CALCIUM ASCORBATE	005743271	182.3189
CALCIUM BENZOATE	002090053	166.110
CALCIUM BROMATE	010102757	136.110 136.115 136.130 136.160 136.180
CALCIUM CAPRATE	013747303	172.863
CALCIUM CAPRYLATE	006107568	172.863
CALCIUM CARBONATE	000471341	184.1191 137.105 182.5191 137.350 169.115 137.155 137.165 137.160 137.170 137.175 137.180 137.185
CALCIUM CASEINATE	009005430	135.140 135.110
CALCIUM CHLORIDE	010035048	145.145 155.200 150.161 PART 133 150.141 172.560 182.6193 182.90 184.1193
CALCIUM CITRATE	000813945	182.5195

MAINTERM	CAS	REGNUM
		150.141
		150.161
		133.173
		155.200
		133.179
		182.6195
		182.1195
		133.169
		182.8195
		184.1195
CALCIUM CYCLAMATE--PROHIBITED	005897165	189.135
CALCIUM DIGLUTAMATE	005996225	
CALCIUM DIPHOSPHATE	007790763	182.5223
		182.8223
CALCIUM FUMARATE	007718516	172.350
CALCIUM GLUCONATE	000299285	150.141
		150.161
		184.1199
CALCIUM GLYCEROPHOSPHATE	027214002	182.5201
		182.8201
CALCIUM HEXAMETAPHOSPHATE	977007694	182.6203
CALCIUM HYDROXIDE	001305620	135.110
		184.1205
CALCIUM HYPOPHOSPHITE	007789799	
CALCIUM IODATE	007789802	136.110
		184.1206
		582.80
		136.115
		136.130
		136.160
		136.180
CALCIUM LACTATE	000814802	184.1207
		145.145
		150.141
		150.161
		155.200
CALCIUM LACTOBIONATE	005001514	172.720
CALCIUM LAURATE	004696564	172.863
CALCIUM LIGNOSULFATE	008061527	172.715
CALCIUM MYRISTATE	015284512	172.863
CALCIUM OLEATE	000142176	172.863
CALCIUM OXIDE	001305788	182.5210

MAINTERM	CAS	REGNUM
		184.1210
CALCIUM PALMITATE	000542427	172.863
CALCIUM PANTOTHENATE	000137086	182.5212
		184.1212
CALCIUM PANTOTHENATE, CALCIUM CHLORIDE DOUBLE SALT	006363388	172.330
CALCIUM PEROXIDE	001305799	136.110
		136.115
		136.130
		136.160
		136.180
CALCIUM PHOSPHATE, DIBASIC	007757939	182.1217
		182.5217
		137.105
		137.155
		182.8217
		137.165
		137.160
		137.170
		137.175
		137.180
		137.185
CALCIUM PHOSPHATE, MONOBASIC	007758238	182.5217
		182.6215
		182.1217
		137.175
		150.161
		137.180
		150.141
		136.110
		155.200
		137.165
		137.270
		136.115
		136.130
		136.160
		136.165
		136.180
		182.8217
CALCIUM PHOSPHATE, TRIBASIC	007758874	169.179
		137.105
		182.1217
		182.5217
		137.155
		182.8217
		137.165
		137.160
		137.170
		137.175
		137.180
		137.185
		169.182
CALCIUM PHYTATE	007776285	
CALCIUM PROPIONATE	004075814	184.1221
		133.173
		150.141
		133.169
		133.123

MAINTERM	CAS	REGNUM
		133.124
		150.161
		133.179
		136.110
		136.115
		136.130
		136.160
		136.180
CALCIUM SALTS OF FATTY ACIDS	977089534	172.863
CALCIUM SILICATE	001344952	172.410
		182.2227
		169.179
		133.146
		182.2906
		573.260
		169.182
CALCIUM SORBATE	007492559	166.110
		182.3225
CALCIUM STEARATE	001592230	173.340
		169.179
		172.863
		169.182
		184.1229
CALCIUM STEAROYL-2-LACTYLATE	005793942	172.844
CALCIUM SULFATE	007778189	155.200
		184.1230
		150.161
		150.141
		137.105
		133.165
		133.195
		133.141
		133.181
		133.183
		133.111
		133.106
		133.102
		137.155
		137.165
		137.160
		137.170
		137.175
		137.180
		137.185
CALUMBA ROOT, EXTRACT (JATRORRHIZA PALMATA (LAM.) MIERS)	977000740	172.510
CALUMBA ROOT(JATRORRHIZA PALMATA (LAM.) MIERS)	977000557	172.510
CAMPHENE	000079925	172.515
CAMPHOLENE ACETATE	001727680	
A&-CAMPHOLENIC ALCOHOL	001901388	
D-CAMPHOR	000464493	172.515

MAINTERM	CAS	REGNUM
CAMPHOR, JAPANESE, WHITE, OIL (CINNAMOMUM CAMPHORA (L.) NEES ET EBERM.)	008008513	172.510
CAMPHOR OIL, FORMOSAN HO-SHO, LEAVES (CINNAMOMUM CAMPHORA SEIB.)	888284568	
CANANGA, OIL (CANANGA ODORATA HOOK. F. AND THOMS.)	068606837	182.20
CANDELILLA WAX (WAX FROM STEMS AND BRANCHES OF EUPHORBIA CERIFERA)	008006448	184.1976
CANDIDA GUILLIERMONDII FOR FERMENTATION PRODUCTION OF CITRIC ACID	977075470	173.160
CANDIDA LIPOLYTICA FOR FERMENTATION PRODUCTION OF CITRIC ACID	068583017	173.165
CANTHAXANTHIN	000514783	73.1075 73.75
CAPERS (CAPPARIS SPINOSA L.)	977050251	182.10
CAPSICUM (CAPSICUM SPP.)	977007729	73.340 182.10
CAPSICUM EXTRACT (CAPSICUM SPP.)	977018420	182.20
CAPSICUM, OLEORESIN (CAPSICUM SPP.)	008023776	73.345 182.20
CARAMEL	008028895	73.85 182.1235 582.1235
CARAWAY, BLACK (NIGELLA SATIVA L.)	977017847	182.10
CARAWAY (CARUM CARVI L.)	977001276	182.10
CARAWAY, OIL (CARUM CARVI L.)	008000428	182.20
CARBOHYDRASE AND CELLULASE FROM ASPERGILLUS NIGER	977050273	173.120
CARBOHYDRASE AND PROTEASE, MIXTURE, FROM BACILLUS SUBTILIS	977082975	
CARBOHYDRASE FROM ASPERGILLUS ORYZAE	977017325	
CARBOHYDRASE FROM BACILLUS LICHENIFORMIS	977043278	

MAINTERM	CAS	REGNUM
CARBOHYDRASE FROM RHIZOPUS ORYZAE	977050284	173.130
CARBOHYDRASE FROM SACCHAROMYCES SPP.	977031723	
CARBON DIOXIDE	000124389	184.1240 169.115 169.140 169.150
CARBOXYMETHYL CELLULOSE	009000117	182.70
CARBOXYMETHYL CELLULOSE, SODIUM SALT	009004324	173.310 182.1745 150.161 133.179 133.134 182.70 150.141 133.178
CARBOXYMETHYL HYDROXYETHYL CELLULOSE	009004302	
CARDAMOM (ELLETARIA CARDAMOMUM (L.) MATON)	977005950	182.10
CARDAMOM OLEORESIN	977090826	182.20
CARDAMOM SEED, OIL (ELLETARIA CARDAMOMUM (L.) MATON)	008000666	182.20
CARMINE (COCCUS CACTI L.)	001390654	73.100 73.1100
CARNAUBA WAX (COPERNICIA CEREFERIA (ARRUDA) MART.)	008015869	184.1978
L-CARNITINE	000541151	
CAROB BEAN, EXTRACT (CERATONIA SILIQUA L.)	977081869	182.20
CAROB BEAN GUM (CERATONIA SILIQUA L.)	009000402	184.1343 133.178 150.141 133.133 133.162 150.161 133.134 133.179 186.1343 240.1051
B&-CAROTENE	007235407	73.95 166.110 182.5245 184.1245

MAINTERM	CAS	REGNUM
CARRAGEENAN	009000071	182.7255
		172.620
		133.134
		133.162
		150.161
		133.178
		133.133
		139.122
		150.141
		136.110
		139.121
		133.179
		136.115
		136.130
		136.160
		136.180
CARRAGEENAN, AMMONIUM SALT OF	060063903	150.141
		139.122
		139.121
		136.110
		150.161
		172.626
		136.115
		136.130
		136.160
		136.180
CARRAGEENAN, AMMONIUM SALT OF, WITH POLYSORBATE 80	977089283	172.623
CARRAGEENAN AND SALTS OF CARRAGEENAN	977043698	172.626
		172.620
CARRAGEENAN, CALCIUM SALT OF	009049052	172.626
		136.110
		150.161
		150.141
		139.122
		139.121
		136.115
		136.130
		136.160
		136.180
CARRAGEENAN, CALCIUM SALT OF, WITH POLYSORBATE 80	977089294	172.623
CARRAGEENAN, POTASSIUM SALT OF	064366241	150.161
		172.626
		150.141
		136.110
		139.122
		139.121
		136.115
		136.130
		136.160
		136.180
CARRAGEENAN, POTASSIUM SALT OF, WITH POLYSORBATE 80	977089307	172.623
CARRAGEENAN SALTS WITH POLYSORBATE 80	977043701	172.623
CARRAGEENAN, SODIUM SALT OF	009061829	172.626

MAINTERM	CAS	REGNUM
		139.122
		136.110
		150.161
		139.121
		150.141
		136.115
		136.130
		136.160
		136.180
CARRAGEENAN, SODIUM SALT OF, WITH POLYSORBATE 80	977089318	172.623
CARRAGEENAN WITH POLYSORBATE 80	977043687	172.623
CARROT, OIL (DAUCUS CAROTA L.)	008015881	182.20 73.300
CARVACROL	000499752	172.515
CARVACRYL ETHYL ETHER	004732132	172.515
CARVEOL	000099489	172.515
4-CARVOMENTHENOL	000562743	172.515
CARVOMENTHOL	000499694	
CARVONE	000099490	182.60
CIS-CARVONE OXIDE	018383498	172.515
CARVYL ACETATE	000097427	172.515
CARVYL PROPIONATE	000097450	172.515
B&-CARYOPHYLLENE	000087445	172.515
B&-CARYOPHYLLENE ALCOHOL	000472979	
CARYOPHYLLENE ALCOHOL	004586225	172.515
B&-CARYOPHYLLENE ALCOHOL ACETATE	062532518	
CARYOPHYLLENE ALCOHOL ACETATE	017622354	172.515
B&-CARYOPHYLLENE OXIDE	001139306	172.515
CASCARA, BITTERLESS, EXTRACT (RHAMNUS PURSHIANA DC.)	977090837	172.510
CASCARILLA BARK, EXTRACT (CROTON SPP.)	977083536	182.20
CASCARILLA BARK, OIL (CROTON SPP.)	008007065	182.20

MAINTERM	CAS	REGNUM
CASEIN	009000719	182.90 166.110
CASSIE, ABSOLUTE (ACACIA FARNESIANA (L.) WILLD.)	977017585	172.510
CASTOREUM, EXTRACT (CASTOR SPP.)	008023834	182.50
CASTOREUM, LIQUID (CASTOR SPP.)	977016899	182.50
CASTOR OIL(RICINUS COMMUNIS L.)	008001794	172.510 177.2600 176.210 73.1 178.3910 178.3120 172.876 177.2800 175.105
CATALASE FROM ASPERGILLUS NIGER	977031847	
CATALASE FROM BOVINE LIVER	081457956	133.113 24.246
CATALASE FROM PENICILLIUM NOTATUM	977090053	
CATECHU, EXTRACT (ACACIA CATECHU WILLD.)	008001761	172.510
CATECHU, POWDER (ACACIA CATECHU WILLD.)	977090848	172.510
CEDAR LEAF, OIL (THUJA OCCIDENTALIS L.)	008007203	172.510
CEDARWOOD OIL ALCOHOLS	068603225	172.515
CEDARWOOD OIL TERPENES	068608322	172.515
CEDRYL ACETATE	000077543	
CELERY SEED (APIUM GRAVEOLENS L.)	977007752	182.10
CELERY SEED, EXTRACT (APIUM GRAVEOLENS L.)	089997353	182.20
CELERY SEED, EXTRACT SOLID (APIUM GRAVEOLENS L.)	977038531	182.20
CELERY SEED, OIL (APIUM GRAVEOLENS L.)	008015905	182.20
CELERY SEED, OLEORESIN	977090860	182.20

MAINTERM	CAS	REGNUM
CELLULOSE ACETATE	009004357	182.90
CELLULOSE, ETHYL	009004573	182.90 73.1 172.868 573.420
CELLULOSE, METHYL	009004675	182.1480 150.141 150.161
CELLULOSE, METHYL ETHYL	009004595	172.872
CELLULOSE, MICROCRYSTALLINE	977005289	133.146
CENTAURY (CENTAURIUM UMBELLATUM GILIB.)	977052779	172.510
CEREAL SOLIDS, HYDROLYZED	977104985	
CHAMOMILE FLOWER (ANTHEMIS NOBILIS L.)	977007263	182.10
CHAMOMILE FLOWER, ENGLISH, OIL (ANTHEMIS NOBILIS L.)	008015927	182.20
CHAMOMILE FLOWER, HUNGARIAN, OIL (MATRICARIA CHAMOMILLA L.)	008002662	182.20
CHAMOMILE FLOWER (MATRICARIA CHAMOMILLA L.)	977001969	182.10
CHAMOMILE FLOWER, ROMAN, EXTRACT (ANTHEMIS NOBILIS L.)	977060119	182.20
CHAR SMOKE FLAVOR	977102149	133.181
CHERRY BARK, WILD, EXTRACT (PRUNUS SEROTINA EHRH.)	977071536	182.20
CHERRY-LAUREL LEAVES (PRUNUS LAUROCERASUS L.)	977052780	172.510
CHERRY LAUREL, OIL (PRUNUS LAUROCERASUS L.) (FFPA)	008000440	
CHERRY-LAUREL WATER (PRUNUS LAUROCERASUS L.)	977038564	172.510
CHERRY PITS, EXTRACT (PRUNUS SPP.)	977038542	172.510
CHERVIL (ANTHRISCUS CEREFOLIUM (L.) HOFFM.)	001338803	182.10
CHERVIL, EXTRACT (ANTHRISCUS CEREFOLIUM L.)	085085207	182.20

MAINTERM	CAS	REGNUM
CHESTNUT LEAVES (CASTANEA DENTATA (MARSH.) BORKH.)	977052791	172.510
CHESTNUT LEAVES, EXTRACT (CASTANEA DENTATA (MARSH.) BORKH.)	977023214	172.510
CHESTNUT LEAVES, EXTRACT SOLID (CASTANEA DENTATA (MARSH.) BORKH.)	977038586	172.510
CHICLE (MANILKARA ZAPOTILLA GILLY AND M. CHICLE GILLY)	008021770	172.615
CHICLE, VENEZUELAN (MANILKARA WILLIAMSII STANDLEY AND RELATED SPP.)	977081427	172.615
CHICORY, EXTRACT (CICHORIUM INTYBUS L.)	068650431	182.20
CHILTE (CNIDOSCOLUS (ALSO KNOWN AS JATROPHA) SPP.)	977011407	172.615
CHIQUIBUL (MANILKARA ZAPOTILLA GILLY)	977052586	172.615
CHIRATA, EXTRACT (SWERTIA CHIRATA BUCH.-HAM.)	977091216	172.510
CHIRATA (SWERTIA CHIRATA BUCH.-HAM.)	977052804	172.510
CHIVES (ALLIUM SCHOENOPRASUM L.)	977050375	182.10
CHLORINE	007782505	137.200 137.105 137.205 137.155 137.165 137.160 137.170 137.175 137.180 137.185
CHLORINE DIOXIDE	010049044	137.200 137.105 137.205 137.155 137.165 137.160 137.170 137.175 137.180 137.185
CHLORINE SOLUTION, AQUEOUS	977091227	
CHLOROFLUOROCARBON 113	000076131	173.342
CHLOROFLUOROCARBONS--PROHIBITED	977084302	189.191

MAINTERM	CAS	REGNUM
CHLOROFORM	000067663	177.1580 175.105
CHLOROMETHYL METHYL ETHER	000107302	173.20
CHLOROPENTAFLUOROETHANE	000076153	173.345
CHLOROPHYLL	001406651	
CHOLIC ACID	000081254	
CHOLINE BITARTRATE	000087672	182.5250 182.8250
CHOLINE CHLORIDE	000067481	182.5252 182.8252
CINCHONA BARK, RED (CINCHONA SUCCIRUBRA PAV. OR ITS HYBRIDS)	977052815	172.510
CINCHONA BARK, RED, EXTRACT (CINCHONA SUCCIRUBRA PAV. OR ITS HYBRIDS)	977038611	172.510
CINCHONA BARK, YELLOW (CINCHONA SPP.)	977052826	172.510
CINCHONA BARK, YELLOW, EXTRACT (CINCHONA SPP.)	977083241	172.510
CINCHONA, EXTRACT (CINCHONA SPP.)	068990125	172.510
1,4-CINEOLE	000470677	172.515
CINNAMALDEHYDE	000104552	182.60
CINNAMALDEHYDE ETHYLENE GLYCOL ACETAL	005660606	172.515
CINNAMIC ACID	000621829	172.515
CINNAMON BARK, EXTRACT (CINNAMOMUM SPP.)	977038600	182.20
CINNAMON BARK, OIL (CINNAMOMUM SPP.)	008007805	182.20 145.140 145.145 145.135 145.180 PART 145 145.181
CINNAMON BARK OLEORESIN, CEYLON, CHINESE, OR SAIGON (CINNAMOMUM SPP.)	977091238	182.20

MAINTERM	CAS	REGNUM
CINNAMON (CINNAMOMUM SPP.)	977000660	182.10
CINNAMON FLOWERS (CINNAMOMUM CASSIA BLUME)	977091249	182.20
CINNAMON LEAF, OIL (CINNAMOMUM SPP.)	008015961	182.20
CINNAMYL ACETATE	000103548	172.515
CINNAMYL ALCOHOL	000104541	172.515
CINNAMYL ANTHRANILATE -- PROHIBITED	000087296	189.113
CINNAMYL BENZOATE	005320752	172.515
CINNAMYL BUTYRATE	000103617	172.515
CINNAMYL CINNAMATE	000122690	172.515
CINNAMYL FORMATE	000104654	172.515
CINNAMYL ISOBUTYRATE	000103593	172.515
CINNAMYL ISOVALERATE	000140272	172.515
CINNAMYL PHENYLACETATE	007492651	172.515
CINNAMYL PROPIONATE	000103560	172.515
CITRAL	005392405	182.60
CITRAL DIETHYL ACETAL	007492662	172.515
CITRAL DIMETHYL ACETAL	007549373	172.515
CITRAL PROPYLENE GLYCOL ACETAL	010444505	172.515
CITRIC ACID	000077929	173.165
		172.755
		182.6033
		182.1033
		PART 133
		PART 146
		161.190
		PART 169
		PART 150
		155.130
		145.145
		131.111
		131.112
		131.136
		131.144
		131.138
		131.146
		146.187

MAINTERM	CAS	REGNUM
		150.161
		150.141
		166.40
		169.115
		169.140
		169.150
		173.160
		173.280
		145.131
		166.110
		184.1033
CITRONELLAL	000106230	172.515
CITRONELLA, OIL (CYMBOPOGON NARDUS RENDLE)	008000291	182.20
DL-CITRONELLOL	026489010	172.515
CITRONELLOXYACETALDEHYDE	007492673	172.515
CITRONELLYL ACETATE	000150845	172.515
CITRONELLYL BUTYRATE	000141162	172.515
CITRONELLYL FORMATE	000105851	172.515
CITRONELLYL ISOBUTYRATE	000097892	172.515
CITRONELLYL PHENYLACETATE	000139708	172.515
CITRONELLYL PROPIONATE	000141140	172.515
CITRONELLYL VALERATE	007540536	172.515
CITRUS PEELS, EXTRACT (CITRUS SPP.)	977038622	182.20
CITRUS RED NO. 2	006358538	74.302
CIVET, ABSOLUTE (VIVERRA CIVETTA SCHREBER AND VIVERRA ZIBETHA SCHREBER)	068916267	182.50
CLARY, OIL (SALVIA SCLAREA L.)	008016635	182.20
CLARY (SALVIA SCLAREA L.)	977051947	182.10
CLAY, ATTAPULGITE	012174117	182.99 582.99
CLOVE BUD, EXTRACT (EUGENIA SPP.)	084961502	184.1257
CLOVE BUD, OIL (EUGENIA SPP.)	008000348	184.1257
CLOVE BUD, OLEORESIN (EUGENIA SPP.)	977017858	184.1257

MAINTERM	CAS	REGNUM
CLOVE LEAF, OIL (EUGENIA SPP.)	008015972	184.1257
CLOVER, EXTRACT(TRIFOLIUM SPP.)	977070511	182.20
CLOVER, OIL(TRIFOLIUM SPP.)	977042184	182.20
CLOVER TOPS, RED, EXTRACT SOLID (TRIFOLIUM PRATENSE L.)	977038655	182.20
CLOVER (TRIFOLIUM SPP.)	977002837	182.10
CLOVES (EUGENIA SPP.)	977007796	184.1257 145.180 145.140 PART 145 145.181
CLOVE STEM, OIL (EUGENIA SPP.)	008015983	184.1257
CLOVE SWEET (MELILOTUS COERULEA)	888284784	
COBALT SULFATE--PROHIBITED	010124433	189.120
COCA LEAF, EXTRACT (DECOCAINIZED) (ERYTHROXYLON COCA LAM.)	977073623	182.20
COCHINEAL EXTRACT (COCCUS CACTI L.)	001260179	73.1100 73.100
COCOA BUTTER SUBSTITUTE FROM COCONUT OIL, PALM KERNEL OIL OR BOTHOILS	085665334	172.861
COCOA BUTTER SUBSTITUTE FROM PALM OIL	002190274	184.1259
COCOA EXTRACT	977075458	182.20 163.113
COCOA WITH DIOCTYL SODIUM SULFOSUCCINATE	977038677	172.520 163.117
COCONUT OIL	008001318	
COCONUT OIL, REFINED	977082986	182.70
COFFEE CONCENTRATE, PURE	977091250	182.20
COFFEE EXTRACT (COFFEA SPP.)	977011827	182.20
COFFEE EXTRACT, SOLID	977091261	182.20
COGNAC, GREEN, OIL	008016215	182.50

MAINTERM	CAS	REGNUM
COGNAC, WHITE, OIL	977050499	182.50
COLLAGEN, AVITENE	009007345	
COMBUSTION PRODUCT GAS	977054264	173.350
COPAIBA, OIL (SOUTH AMERICAN SPP. OF COPAIFERA L.)	008013976	172.510
COPAIBA (SOUTH AMERICAN SPP. OF COPAIFERA L.)	008001614	172.510
COPALS, MANILA	009000424	73.1
COPPER GLUCONATE	000527093	582.80 182.5260 184.1260
COPPER SULFATE	007758987	582.80 184.1261
CORIANDER (CORIANDRUM SATIVUM L.)	977007810	182.10
CORIANDER, OIL (CORIANDRUM SATIVUM L.)	008008524	182.20
CORK, OAK (QUERCUS SPP.)	977038688	172.510
CORN ENDOSPERM OIL	977010506	73.315
CORN GLUTEN	066071963	184.1321
CORN MINT OIL	068917180	
CORN SILK EXTRACT (ZEA MAYS L.)	977071116	184.1262
CORN SILK, OIL(ZEA MAYS L.)	977089409	184.1262
CORN SYRUP	977004128	184.1865 133.124 PART 146 155.200 163.123 133.178 155.194 133.179 145.134 PART 145 169.175 169.179 163.150 163.153 169.176 169.177 169.178 169.180 169.181 169.182

MAINTERM	CAS	REGNUM
COSTMARY (CHRYSANTHEMUM BALSAMITA L.)	977017869	172.510
COSTUS ROOT, OIL (SAUSSUREA LAPPA CLARKE)	008023889	172.510
COTTONSEED FLOUR, DEFATTED	977100176	172.894
COTTONSEED FLOUR, PARTIALLY DEFATTED, COOKED	977050546	172.894
COTTONSEED FLOUR, PARTIALLY DEFATTED, COOKED, TOASTED	977043778	73.140 172.894
COTTONSEED KERNELS, GLANDLESS, RAW	977043563	172.894
COTTONSEED KERNELS, GLANDLESS, ROASTED	977043789	172.894
COUMARIN--PROHIBITED	000091645	189.130
COUMARONE-INDENE RESINS	063393895	172.215
M-CRESOL	000108394	
O-CRESOL	000095487	
P-CRESOL	000106445	172.515
CUBEBS, OIL (PIPER CUBEBA L. F.)	008007872	172.510
CUBEBS (PIPER CUBEBA L. F.)	977000820	172.510
CUMINALDEHYDE	000122032	172.515
CUMIN (CUMINUM CYMINUM L.)	977050557	182.10
CUMIN, OIL (CUMINUM CYMINUM L.)	008014139	182.20
CUPROUS IODIDE	007681654	582.80 184.1265
CURRANT BUDS, BLACK, ABSOLUTE (RIBES NIGRUM L.)	068606815	172.510
CURRANT JUICE, BLACK	977038702	172.510
CURRANT LEAVES, BLACK (RIBES NIGRUM L.)	977032419	172.510
CYCLAMATE--PROHIBITED	977016968	189.135

MAINTERM	CAS	REGNUM
CYCLOHEPTADECA-9-EN-1-ONE	000542461	
CYCLOHEXANE	000110827	73.1 175.105 176.200
CYCLOHEXANEACETIC ACID	005292217	172.515
CYCLOHEXANECARBOXYLIC ACID	000098895	
CYCLOHEXANEETHYL ACETATE	021722838	172.515
CYCLOHEXYL ACETATE	000622457	172.515
CYCLOHEXYLAMINE	000108918	173.310
CYCLOHEXYL ANTHRANILATE	007779160	172.515
CYCLOHEXYL BUTYRATE	001551446	172.515
CYCLOHEXYL CINNAMATE	007779171	172.515
CYCLOHEXYL FORMATE	004351546	172.515
CYCLOHEXYL ISOVALERATE	007774449	172.515
CYCLOHEXYLMETHYL PYRAZINE	028217927	
CYCLOHEXYL PROPIONATE	006222351	172.515
CYCLOPENTANETHIOL	001679078	
P-CYMENE	000099876	172.515
L-CYSTEINE	000052904	172.320 184.1271
L-CYSTEINE MONOHYDROCHLORIDE	000052891	184.1272
DL-CYSTINE	000923320	
L-CYSTINE	000056893	172.320
DAMAR GUM (SHOREA DIPTEROCARPACEAE)	009000162	73.1 177.1400 177.1200
D&-DAMASCONE	057378684	
A&-DAMASCONE	043052875	
DAMIANA LEAVES (TURNERA DIFFUSA WILLD.)	977000853	172.510

MAINTERM	CAS	REGNUM
DANDELION, FLUID EXTRACT (TARAXACUM SPP.)	977038713	182.20
DANDELION ROOT, EXTRACT SOLID (TARAXACUM SPP.)	977038724	182.20
DAVANA OIL (ARTEMESIA PALLENS WALL.)	008016033	172.510
2-TRANS,4-TRANS-DECADIENAL	025152845	
D&-DECALACTONE	000705862	172.515
E&-DECALACTONE	005579782	
G&-DECALACTONE	000706149	172.515
DECANAL	000112312	182.60
DECANAL DIMETHYL ACETAL	007779411	172.515
DECANOIC ACID	000334485	172.860 173.340 172.210
1-DECANOL	000112301	172.864 172.515
3-DECANOL	001565817	
2-DECENAL	003913711	172.515
4-DECENAL	030390502	
5-DECENOIC ACID	085392036	
6-DECENOIC ACID	085392047	
9-DECENOIC ACID	014436329	
3-DECEN-2-ONE	010519332	172.515
DECYL ACETATE	000112174	172.515
DECYL BUTYRATE	005454091	172.515
DECYL PROPIONATE	005454193	172.515
DEERTONGUE SOLID EXTRACT	068602868	
DEHYDRATED BEETS	977010482	73.40

MAINTERM	CAS	REGNUM
DEHYDROACETIC ACID	000520456	172.130
DEHYDRODIHYDROIONOL	057069860	
DEHYDRODIHYDROIONONE	020483367	
DEHYDROMENTHOFUROLACTONE	075640265	
DESOXYCHOLIC ACID	000083443	
DEXTRANS (AVG M W LESS THAN 100,000)	009004540	186.1275
DEXTRIN	009004539	184.1277
DEXTROSE	000050997	163.123 133.124 PART 146 133.178 133.179 155.194 169.175 145.134 155.200 155.170 169.179 145.180 PART 145 145.181 163.150 163.153 169.176 169.177 169.178 169.180 169.181 169.182 184.1857
DIACETYL	000431038	184.1278
DIALKANOLAMIDE	977046313	173.315(A)(3) 173.315
DI-N-ALKYL(C8-C18 FROM COCONUT OIL) DIMETHYL AMMONIUM CHLORIDE	061789773	172.710
DIALLYL POLYSULFIDES	072869751	
DIALLYL TRISULFIDE	002050875	
DIASTASE FROM ASPERGILLUS ORYZAE	009001198	
DIATOMACEOUS EARTH	061790532	573.340 182.90
DIBENZYL ETHER	000103504	172.515
2,2-DIBROMO-3-NITRILOPROPIONAMIDE	010222012	173.320

MAINTERM	CAS	REGNUM
DI-(BUTAN-3-ONE-1-YL) SULFIDE	040790043	
4,4-DIBUTYL-G&-BUTYROLACTONE	007774472	172.515
DIBUTYL SEBACATE	000109433	172.515
DICHLORODIFLUOROMETHANE	000075718	173.355
DICYCLOHEXYL DISULFIDE	002550405	
DIETHANOLAMIDE CONDENSATE FROM SOYBEAN OIL FATTY ACIDS(C16-C18)	068425478	172.710
DIETHANOLAMIDE CONDENSATE FROM STRIPPED COCONUT OIL FATTY ACIDS(C10-C18)	977103824	172.710
1,2-(DI(1'-ETHOXY)ETHOXY)PROPANE	067715791	
DIETHYLAMINOETHANOL	000100378	173.310
DIETHYLENE GLYCOL DISTEARATE	000109308	73.1
DIETHYLENETRIAMINE	000111400	173.20
DIETHYLENETRIAMINE CROSSLINKED WITH EPICHLOROHYDRIN	025085170	173.25
DI(2-ETHYLHEXYL) ADIPATE	000103231	175.105 175.300 177.1200 177.1210 177.1400 178.3740
DI(2-ETHYLHEXYL) PHTHALATE	000117817	175.105 175.310 175.380 175.390 176.170 177.1010 177.1200 177.1210 177.1400 178.3120 178.3910 181.27
DIETHYL MALATE	007554123	172.515
DIETHYL MALONATE	000105533	172.515
2,3-DIETHYL-5-METHYLPYRAZINE	018138040	
2,3-DIETHYLPYRAZINE	015707241	

MAINTERM	CAS	REGNUM
DIETHYL PYROCARBONATE -- PROHIBITED	001609478	189.140
DIETHYL SEBACATE	000110407	172.515
DIETHYL SUCCINATE	000123251	172.515
DIETHYL TARTRATE	000087912	172.515
2,5-DIETHYLTETRAHYDROFURAN	041239489	172.515
DIFURFURYL ETHER	004437223	
DIHYDROCARVEOL	000619012	172.515
CIS-DIHYDROCARVONE	003792538	172.515
DIHYDROCARVYL ACETATE	020777495	172.515
DIHYDROCOUMARIN	000119846	
4,5-DIHYDRO-3(2H)THIOPHENONE	001003049	
DIHYDRO-B&-IONOL	003293478	
DIHYDRO-A&-IONONE	031499726	
DIHYDRO-B&-IONONE	017283817	
3,6-DIHYDRO-4-METHYL-2(2-METHYLPRO-PEN-1-YL)-2H-PYRAN	001786089	
5,7-DIHYDRO-2-METHYLTHIENO(3,4-D)P-YRIMIDINE	036267717	
DIHYDROXYACETOPHENONE	028631869	
DILAURYL THIODIPROPIONATE	000123284	182.3280
DILL (ANETHUM GRAVEOLENS L.)	977050604	184.1282
DILL, OIL (ANETHUM GRAVEOLENS L.)	008006755	184.1282
DILL SEED, INDIAN (ANETHUM SPP.)	977082997	172.510 184.1282
M-DIMETHOXYBENZENE	000151100	172.515
P-DIMETHOXYBENZENE	000150787	172.515
1,1-DIMETHOXYETHANE	000534156	
2,6-DIMETHOXYPHENOL	000091101	

MAINTERM	CAS	REGNUM
3,4-DIMETHOXY-1-VINYLBENZENE	006380230	
2,4-DIMETHYLACETOPHENONE	000089747	172.515
1,4-DIMETHYL-4-ACETYL-1-CYCLOHEXENE	043219687	
2,4-DIMETHYL-5-ACETYLTHIAZOLE	038205606	
DIMETHYLAMINE	000124403	173.20
2,4-DIMETHYLBENZALDEHYDE	015764166	
2,3-DIMETHYLBENZOFURAN	003782001	
P,A&-DIMETHYLBENZYL ALCOHOL	000536505	
A&,A&-DIMETHYLBENZYL ISOBUTYRATE	007774609	172.515
3,4-DIMETHYL-1,2-CYCLOPENTADIONE	013494069	
3,5-DIMETHYL-1,2-CYCLOPENTADIONE	013494070	
DIMETHYL DIALKYL AMMONIUM CHLORIDE	977065954	173.400
DIMETHYL DICARBONATE	004525331	172.133
2,5-DIMETHYL-2,5-DIHYDROXY-1,4-DIT-HIANE	055704784	
DIMETHYLETHANOLAMINE	000108010	173.20
4,5-DIMETHYL-2-ETHYL-3-THIAZOLINE	076788460	
2,5-DIMETHYL-3-FURANTHIOL	055764233	
2,6-DIMETHYL-4-HEPTANOL	000108827	
2,6-DIMETHYL-4-HEPTANONE	000108838	
2,6-DIMETHYL-5-HEPTENAL	000106729	172.515
2,6-DIMETHYL-6-HEPTEN-1-OL	040326010	
4,5-DIMETHYL-3-HYDROXY-2,5-DIHYDRO-FURAN-2-ONE	028664359	
4,5-DIMETHYL-2-ISOBUTYL-3-THIAZOLI-NE	065894839	
2,5-DIMETHYL-4-METHOXY-3(2H)-FURAN-ONE	004077478	

MAINTERM	CAS	REGNUM
2,6-DIMETHYL-10-METHYLENE-2,6,11-D-ODECATRIENAL	060066888	
2,6-DIMETHYL-3-((2-METHYL-3-FURYL)-THIO)-4-HEPTANONE	061295510	
2,2-DIMETHYL-5-(1-METHYLPROPEN-1-YL) TETRAHYDROFURAN	007416355	
2-TRANS-3,7-DIMETHYLOCTA-2,6-DIENYL 2-ETHYLBUTANOATE	073019144	
2,6-DIMETHYLOCTANAL	007779079	172.515
3,7-DIMETHYL-1-OCTANOL	000106218	172.515
3,7-DIMETHYL-6-OCTENOIC ACID	000502476	
2,4-DIMETHYL-2-PENTENOIC ACID	066634977	
A&,A&-DIMETHYLPHENETHYL ACETATE	000151053	172.515
A&,A&-DIMETHYLPHENETHYL ALCOHOL	000100867	172.515
A&,A&-DIMETHYLPHENETHYL BUTYRATE	010094345	172.515
A&,A&-DIMETHYLPHENETHYL FORMATE	010058432	172.515
DIMETHYLPOLYSILOXANE	009006659	173.340 146.185 145.180 145.181
2,3-DIMETHYLPYRAZINE	005910894	
2,5-DIMETHYLPYRAZINE	000123320	
2,6-DIMETHYLPYRAZINE	000108509	
2,6-DIMETHYLPYRIDINE	000108485	
2,5-DIMETHYLPYRROLE	000625843	
P,A&-DIMETHYLSTYRENE	001195320	
DIMETHYL SUCCINATE	000106650	172.515
4,5-DIMETHYLTHIAZOLE	003581917	
2,5-DIMETHYL-3-THIOISOVALERYLFURAN	055764288	
2,6-DIMETHYLTHIOPHENOL	000118729	

MAINTERM	CAS	REGNUM
DIMETHYL TRISULFIDE	003658808	
3,5-DIMETHYL-1,2,4-TRITHIOLANE	023654924	
2,4-DIMETHYL-5-VINYLTHIAZOLE	065505182	
DIOCTYL SODIUM SULFOSUCCINATE	000577117	172.810 169.150 133.134 73.1 172.520 133.133 169.115 163.114 133.162 133.179 131.130 133.124 131.132 133.178 172.808 163.117
DIPHENYL ETHER	000101848	
1,3-DIPHENYL-2-PROPANONE	000102045	172.515
DIPROPYL TRISULFIDE	006028611	
DISODIUM CYANODITHIOIMIDOCARBONATE	000138932	173.320
DISODIUM ETHYLENEBISDITHIOCARBAMATE	000142596	173.320
DISODIUM GUANYLATE	005550129	172.530 155.120 155.200 155.170 155.130 145.131
DISODIUM INOSINATE	004691650	155.200 172.535 155.170 155.130 155.120 145.131
DISODIUM SUCCINATE	000150903	
2,8-DITHIANON-4-EN-4-CARBOXALDEHYDE	059902011	
2,2'-(DITHIODIMETHYLENE) DIFURAN	004437201	
DITTANY (FRAXINELLA) ROOTS (DICTAMNUS ALBUS L.)	977047656	172.510
DITTANY OF CRETE (ORIGANUM DICTAMNUS L.)	977017927	172.510

MAINTERM	CAS	REGNUM
2-TRANS-6-CIS-DODECADIENAL	021662135	
TRANS,TRANS-2,4-DODECADIENAL	021662168	
D&-DODECALACTONE	000713951	172.515
E&-DODECALACTONE	016429213	
G&-DODECALACTONE	002305057	172.515
2-DODECENAL	004826624	172.515
DODECYL GALLATE	001166525	166.110
DODECYL ISOBUTYRATE	006624711	
DOGGRASS, EXTRACT (AGROPYRON REPENS (L.) BEAUV.)	977038735	182.20
DRAGON'S BLOOD, EXTRACT (DAEMONOROPS SPP. OR OTHER BOTANICAL SOURCES)	009000195	172.510
DRIED ALGAE MEAL	977010471	73.275
DULCIN--PROHIBITED	000150696	189.145
EDTA, CALCIUM DISODIUM	000062339	172.120 169.150 169.140 169.115 166.110 155.200 161.173
EDTA, DISODIUM	000139333	155.200 169.150 169.140 172.135 169.115 573.360
EDTA, DISODIUM IRON	014729896	
EDTA, TETRASODIUM	000064028	173.310 173.315(A)(3) 173.315 175.105 176.210 178.3120
ELDER FLOWERS, EXTRACT(SAMBUCUS CANADENSIS L. OR SAMBUCUS NIGRA L.)	977010391	182.20
ELDER FLOWERS (SAMBUCUS CANADENSIS L. OR SAMBUCUS NIGRA L.)	977002473	182.10

MAINTERM	CAS	REGNUM
ELDER TREE LEAVES (SAMBUCUS NIGRA L.)	977038746	172.510
ELECAMPANE ROOT, EXTRACT (INULA HELENIUM L.)	977091272	172.510
ELECAMPANE ROOT, OIL (INULA HELENIUM L.)	001397837	172.510
ELEMI, GUM (CANARIUM SPP.)	009000753	172.510
ELEMI, OIL (CANARIUM SPP.)	008023890	172.510
ENZYME-MODIFIED FATS	977127788	
ENZYMES, BACTERIAL	977143808	
ENZYMES, PROTEOLYTIC	009001927	
EPICHLOROHYDRIN	000106898	172.892
EPICHLOROHYDRIN CROSSLINKED WITH AMMONIA	028551146	173.25
ERIGERON, OIL (ERIGERON CANADENSIS L.)	008007270	172.510
ERYTHORBIC ACID	000089656	182.3041 145.110
ESTERASE-LIPASE FROM MUCOR MIEHEI	977032088	173.140
ESTRAGOLE	000140670	172.515 182.20
1,2-ETHANEDITHIOL	000540636	
P-ETHOXYBENZALDEHYDE	010031820	172.515
O-(ETHOXYMETHYL)PHENOL	020920836	
ETHOXYQUIN	000091532	172.140
2-ETHOXYTHIAZOLE	015679193	
ETHYL ABIETATE	000631710	
ETHYL ACETATE	000141786	73.1 182.60 172.560 173.228
ETHYL ACETOACETATE	000141979	172.515

MAINTERM	CAS	REGNUM
ETHYL 2-ACETYL-3-PHENYLPROPIONATE	000620791	172.515
1-ETHYL-2-ACETYLPYRROLE	039741418	
ETHYL ACONITATE (MIXED ESTERS)	001321308	172.515
ETHYL ACRYLATE	000140885	172.515
ETHYL ALCOHOL	000064175	169.175 169.3 184.1293 172.560 169.176 169.177 169.178 169.180 169.181 172.340
ETHYL P-ANISATE	000094304	172.515
ETHYL ANTHRANILATE	000087252	172.515
4-ETHYLBENZALDEHYDE	004748781	
ETHYL BENZOATE	000093890	172.515
ETHYL BENZOYLACETATE	000094020	172.515
A&-ETHYL BENZYL BUTYRATE	010031864	172.515
ETHYL BRASSYLATE	000105953	172.515
2-ETHYLBUTYL ACETATE	010031875	172.515
2-ETHYLBUTYRALDEHYDE	000097961	172.515
ETHYL BUTYRATE	000105544	182.60
2-ETHYLBUTYRIC ACID	000088095	172.515
ETHYL CINNAMATE	000103366	172.515
ETHYL CROTONATE	000623701	172.515
ETHYL CYCLOHEXANECARBOXYLATE	003289289	
ETHYL CYCLOHEXANEPROPIONATE	010094367	172.515
ETHYL TRANS-2,CIS-4-DECADIENOATE	003025307	
ETHYL DECANOATE	000110383	172.515
ETHYL TRANS-2-DECENOATE	007367886	

MAINTERM	CAS	REGNUM
ETHYL TRANS-4-DECENOATE	076649166	
4-ETHYL-2,6-DIMETHOXYPHENOL	014059928	
2-ETHYL-4,5-DIMETHYLOXAZOLE	053833300	
2-ETHYL-3,(5 OR 6)-DIMETHYLPYRAZINE	027043056	
3-ETHYL-2,6-DIMETHYLPYRAZINE	013925070	
ETHYL 2,4-DIOXOHEXANOATE	013246521	
ETHYLENE DIAMINE	000107153	173.320
ETHYLENE DICHLORIDE	000107062	173.230 172.710 172.560 173.315 573.440
ETHYLENE GLYCOL DISTEARATE	000627838	73.1
ETHYLENE GLYCOL MONOBUTYL ETHER	000111762	173.315(A)(3) 173.315 175.105 176.210 177.1650 178.1010
ETHYLENE GLYCOL MONOETHYL ETHER	000110805	73.1 175.105
ETHYLENE OXIDE	000075218	172.710 172.808 175.105 176.180 176.210 178.3120 178.3520 193.200
ETHYLENE OXIDE POLYMER	009002908	172.770
ETHYLENE OXIDE POLYMER, ALKYL ADDUCT	977047838	173.315(A)(3) 173.315
ETHYLENE OXIDE POLYMER, ALKYL ADDUCT, PHOSPHATE ESTER	977092231	173.315
ETHYLENE OXIDE/PROPYLENE OXIDE COPOLYMER	009003116	172.808 173.340
ETHYLENE OXIDE/PROPYLENE OXIDE COPOLYMER, ALKYL ADDUCT	977083025	173.315

MAINTERM	CAS	REGNUM
ETHYLENE OXIDE/PROPYLENE OXIDE COPOLYMER, ALKYL ADDUCT, PHOSPHATE ESTER	977083036	173.315
ETHYLENE OXIDE/PROPYLENE OXIDE COPOLYMER (AVG M W 14,000)	977057876	172.808(A)(4) 172.808
ETHYLENE OXIDE/PROPYLENE OXIDE COPOLYMER (AVG M W 9,760 - 13,200)	977057912	172.808(A)(1) 172.808
ETHYLENE OXIDE/PROPYLENE OXIDE COPOLYMER (AVG M W 3,500-4,125)	977057832	172.808(A)(2) 172.808
ETHYLENE OXIDE/PROPYLENE OXIDE COPOLYMER (MIN AVG M W 1,900)	977057638	172.808(A)(3) 172.808
ETHYL N-ETHYLANTHRANILATE	038446218	
ETHYL 2-ETHYL-3-PHENYLPROPANOATE	002983360	
ETHYL FORMATE	000109944	184.1295
2-ETHYLFURAN	003208160	172.515
ETHYL 2-FURANPROPIONATE	010031900	172.515
ETHYL 3-(FURFURYLTHIO) PROPIONATE	094278270	
4-ETHYLGUAIACOL	002785899	172.515
ETHYL HEPTANOATE	000106309	172.515
2-ETHYL-2-HEPTENAL	010031886	172.515
ETHYL HEXANOATE	000123660	172.515
2-ETHYL-1-HEXANOL	000104767	
ETHYL 2-HEXENOATE	001552676	
ETHYL 3-HEXENOATE	002396830	
ETHYL TRANS-2-HEXENOATE	027829727	
1-ETHYLHEXYL TIGLATE	094133923	
ETHYL 3-HYDROXYBUTYRATE	005405414	
3-ETHYL-2-HYDROXY-2-CYCLOPENTEN-1--ONE	021835018	

MAINTERM	CAS	REGNUM
ETHYL 3-HYDROXYHEXANOATE	002305251	
3-ETHYL-2-HYDROXY-4-METHYLCYCLOPEN-T-2-EN-1-ONE	042348129	
5-ETHYL-2-HYDROXY-3-METHYLCYCLOPEN-T-2-EN-1-ONE	053263584	
2-ETHYL-4-HYDROXY-5-METHYL-3(2H)-F-URANONE	027538109	
5-ETHYL-3-HYDROXY-4-METHYL-2(5H)-F-URANONE	000698102	
ETHYL ISOBUTYRATE	000097621	172.515
N-ETHYL-2-ISOPROPYL-5-METHYLCYCLOH-EXANE CARBOXAMIDE	039711790	
ETHYL ISOVALERATE	000108645	172.515
ETHYL LACTATE	000097643	172.515
ETHYL LAURATE	000106332	172.515
ETHYL LEVULINATE	000539888	172.515
ETHYL MALTOL	004940118	172.515
ETHYL 2-MERCAPTOPROPIONATE	019788499	
ETHYL 3-MERCAPTOPROPIONATE	005466068	
ETHYL 2-METHYLBUTYRATE	007452791	172.515
ETHYL 2-METHYL-3,4-PENTADIENOATE	060523219	
ETHYL 2-METHYLPENTANOATE	039255328	
ETHYL 3-METHYLPENTANOATE	005870688	
ETHYL 2-METHYL-3-PENTENOATE	001617238	
ETHYL 2-METHYL-4-PENTENOATE	053399818	172.515
ETHYL METHYLPHENYLGLYCIDATE	000077838	182.60
2-ETHYL-5-METHYLPYRAZINE	013360640	
3-ETHYL-2-METHYLPYRAZINE	015707230	
5-ETHYL-2-METHYLPYRIDINE	000104905	

MAINTERM	CAS	REGNUM
2-ETHYL-4-METHYLTHIAZOLE	015679126	
ETHYL 4-(METHYLTHIO) BUTYRATE	022014488	
ETHYL 3-METHYLTHIOPROPIONATE	013327565	
ETHYL METHYL-P-TOLYLGLYCIDATE	074367978	
2-ETHYL-(3 OR 5 OR 6)-MOP(85%) AND 2-METHYL-(3 OR 5 OR 6)-MOP(13%)	977044475	
ETHYL MYRISTATE	000124061	172.515
ETHYL NITRITE	000109955	172.515
ETHYL NONANOATE	000123295	172.515
ETHYL 2-NONYNOATE	010031922	172.515
ETHYL OCTADECANOATE	000111615	
ETHYL CIS-4,7-OCTADIENOATE	069925333	
ETHYL OCTANOATE	000106321	172.515
ETHYL CIS-4-OCTENOATE	034495711	
ETHYL TRANS-2-OCTENOATE	007367820	
ETHYL OLEATE	000111626	172.515
ETHYL 3-OXOHEXANOATE	003249681	
ETHYL PALMITATE	000628977	
ETHYLPARABEN	000120478	175.105
P-ETHYLPHENOL	000123079	
ETHYL PHENYLACETATE	000101973	172.515
ETHYL 4-PHENYLBUTYRATE	010031933	172.515
ETHYL 3-PHENYLGLYCIDATE	000121391	172.515
ETHYL 3-PHENYLPROPIONATE	002021285	172.515
ETHYL PROPIONATE	000105373	172.515
2-ETHYLPYRAZINE	013925003	

MAINTERM	CAS	REGNUM
3-ETHYLPYRIDINE	000536787	
ETHYL PYRUVATE	000617356	172.515
ETHYL SALICYLATE	000118616	172.515
ETHYL SORBATE	002396841	172.515
ETHYL THIOACETATE	000625605	
2-ETHYLTHIOPHENOL	004500587	
ETHYL TIGLATE	005837785	172.515
ETHYL (P-TOLYLOXY)ACETATE	067028404	
2-ETHYL-1,3,3-TRIMETHYL-2-NORBORNA-NOL	018368917	
ETHYL UNDECANOATE	000627907	172.515
ETHYL 10-UNDECENOATE	000692864	172.515
ETHYL VALERATE	000539822	172.515
ETHYL VANILLIN	000121324	163.112
		163.123
		182.90
		182.60
		163.111
		163.130
		163.113
		163.114
		163.117
		163.135
		163.140
		163.145
		163.155
		163.150
		163.153
EUCALYPTOL	000470826	172.515
EUCALYPTUS, OIL (EUCALYPTUS GLOBULUS LABILLE)	008000484	172.510
EUGENOL	000097530	184.1257
EUGENYL ACETATE	000093287	172.515
EUGENYL BENZOATE	000531260	172.515
EUGENYL FORMATE	010031966	172.515
EUGENYL METHYL ETHER	000093152	172.515

MAINTERM	CAS	REGNUM
FARNESOL	004602840	172.515
FD & C BLUE NO. 1	002650182	74.101 74.1101 82.101
FD & C BLUE NO. 2	000860220	74.102 74.1102 82.102
FD & C BLUE NO. 1, ALUMINUM LAKE	068921426	82.101 176.180 82.51 81.1
FD & C BLUE NO. 2, ALUMINUM LAKE	016521383	81.1 82.51 176.180 82.102
FD & C BLUE NO. 1, CALCIUM LAKE	977011134	82.51 176.180 82.101 81.1
FD & C BLUE NO. 2, CALCIUM LAKE	977011145	176.180 81.1 82.51 82.102
FD & C GREEN NO. 3	002353459	74.203 74.1203 74.2203 82.203
FD & C GREEN NO. 3, ALUMINUM LAKE	977011883	176.180 81.1 82.51 82.203
FD & C GREEN NO. 3, CALCIUM LAKE	977011123	82.51 81.1 176.180 82.203
FD & C RED NO. 3	016423680	74.1303 74.303 82.303 81.27
FD & C RED NO. 40	025956176	74.340 74.1340 74.2340
FD & C RED NO. 3, ALUMINUM LAKE	012227780	82.51 176.180
FD & C RED NO. 40, ALUMINUM LAKE	977068862	74.340 82.51 176.180

MAINTERM	CAS	REGNUM
FD & C RED NO. 3, CALCIUM LAKE	977011167	82.303 82.51 176.180 81.1
FD & C RED NO. 40, CALCIUM LAKE	977011178	82.51 176.180 74.340
FD & C YELLOW NO. 5	001934210	74.705 74.1705 74.2705
FD & C YELLOW NO. 6	002783940	74.706 74.1706 82.706 201.20 74.2706
FD & C YELLOW NO. 5, ALUMINUM LAKE	012225217	82.51 81.1 176.180 82.705
FD & C YELLOW NO. 6, ALUMINUM LAKE	015790075	176.180 82.51 82.706 81.1
FD & C YELLOW NO. 5, CALCIUM LAKE	977011189	81.1 82.51 176.180 82.705
FD & C YELLOW NO. 6, CALCIUM LAKE	977083047	82.706 82.51 81.1 176.180
D-FENCHONE	004695629	172.515
FENCHYL ALCOHOL	001632731	172.515
FENNEL, COMMON (FOENICULUM VULGARE MILL.)	977001130	182.10
FENNEL, SWEET (FOENICULUM VULGARE MILL. VAR. DULCE (D.C.) ALEF.)	977007854	182.20
FENNEL, SWEET, OIL (FOENICULUM VULGARE MILL. VAR. DULCE (D.C.) ALEF.)	008006846	182.20
FENUGREEK, EXTRACT (TRIGONELLA FOENUM-GRAECUM L.)	977064677	182.20
FENUGREEK, OLEORESIN (TRIGONELLA FOENUM-GRAECUM L.)	977018533	182.20
FENUGREEK (TRIGONELLA		

MAINTERM	CAS	REGNUM
FOENUM-GRAECUM L.)	008023903	182.10
FERRIC CHLORIDE	007705080	176.170 182.99 175.105 184.1297
FERRIC CITRATE	003522507	184.1298
FERRIC OXIDE	001309371	522.940
FERRIC PEPTONATE	977089863	
FERRIC PHOSPHATE	010045860	582.80 182.5301 184.1301
FERRIC PYROPHOSPHATE	010058443	582.80 182.5304 184.1304
FERRIC SODIUM PYROPHOSPHATE	010045871	182.5306
FERRIC SULFATE	010028225	582.80 182.99 184.1307
FERROCYANIDE SALTS	888284955	
FERROUS ASCORBATE	024808524	184.1307A
FERROUS CARBONATE	000563713	184.1307B
FERROUS CITRATE	023383111	184.1307C
FERROUS FUMARATE	000141015	172.350 184.1307D
FERROUS GLUCONATE	006047127	73.160 582.80 182.5308 184.1308
FERROUS LACTATE	005905522	182.5311 184.1311
FERROUS PEPTONATE	977089874	
FERROUS SULFATE	007782630	182.5315 184.1315
FICIN	009001336	
FIR NEEDLES AND TWIGS, OIL (ABIES SPP.)	008021281	172.510

MAINTERM	CAS	REGNUM
FISH OIL (HYDROGENATED)	091078954	186.1551
FISH PROTEIN CONCENTRATE, WHOLE	977050739	172.385
FISH PROTEIN ISOLATE	977080388	172.340
FOLIC ACID	000059303	172.345
FORMALDEHYDE	000050000	173.340 573.460 175.105 178.3120
FORMIC ACID	000064186	172.515 186.1316 573.480
2-FORMYL-6,6-DIMETHYLBICYCLO(3.1.1-)HEPT-2-ENE	000564943	
FRUIT JUICE	977010517	73.250 PART 131 131.110 131.143
FULLERS EARTH	008031183	
FUMARIC ACID	000110178	150.141 172.350 146.113 150.161 131.144
FUNGAL HEMICELLULASE	977147946	
FUNGAL PECTINASE	977033809	
2-FURANMETHANETHIOL FORMATE	059020905	
FURCELLERAN	009000219	172.655
FURCELLERAN, AMMONIUM SALT OF	977089772	172.660
FURCELLERAN AND SALTS OF FURCELLERAN	977043654	172.660 172.655
FURCELLERAN, CALCIUM SALT OF	977089783	172.660
FURCELLERAN, POTASSIUM SALT OF	977089794	172.660
FURCELLERAN, SODIUM SALT OF	977089807	172.660
FURFURAL	000098011	175.105
FURFURYL ACETATE	000623176	

MAINTERM	CAS	REGNUM
FURFURYL ALCOHOL	000098000	175.105
FURFURYL BUTYRATE	000623212	
2-FURFURYLIDENEBUTYRALDEHYDE	000770274	
FURFURYL ISOPROPYL SULFIDE	001883789	
FURFURYL MERCAPTAN	000098022	
FURFURYL 3-METHYLBUTANOATE	013678609	
FURFURYL METHYL ETHER	013679464	
FURFURYL METHYL SULFIDE	001438911	
A&-FURFURYL OCTANOATE	039252034	
A&-FURFURYL PENTANOATE	036701016	
FURFURYL PROPIONATE	000623198	
N-FURFURYLPYRROLE	001438944	
FURFURYL THIOACETATE	013678687	
FURFURYL THIOPROPIONATE	059020858	
3(2-FUROYLTHIO)-2,5-DIMETHYLFURAN	055764313	
3-(2-FURYL)ACROLEIN	000623303	
1-(2-FURYL)-1,3-BUTANEDIONE	025790356	
4-(2-FURYL)-3-BUTEN-2-ONE	000623154	
2-FURYL METHYL KETONE	001192627	
(2-FURYL)-2-PROPANONE	006975606	172.515
FUSEL OIL, REFINED	008013750	172.515
A&-GALACTOSIDASE FROM MORTEIRELLA VINACEAE RAFFINOSEUTILIZER	977080399	173.145
GALANGA, GREATER (ALPINIA GALANGA WILLD)	977050773	172.510
GALANGAL ROOT (ALPINIA SPP.)	977038757	182.10

MAINTERM	CAS	REGNUM
GALANGAL ROOT, EXTRACT (ALPINIA SPP.)	977038768	182.20
GALANGAL ROOT, OIL (ALPINIA SPP.)	008024406	182.20
GALBANUM, OIL (FERULA SPP.)	008023914	172.510
GALBANUM, RESIN (FERULA SPP.)	009000242	172.510
GAMBIR (UNCARIA GAMBIR ROXB.)	008001487	172.510
GARLIC	977001812	184.1317
GARLIC, OIL (ALLIUM SATIVUM L.)	008000780	184.1317
GELATIN	009000708	182.70 133.162 133.179 133.178 133.134 133.133
GELLAN GUM	071010521	172.665
GENET, ABSOLUTE (SPARTIUM JUNCEUM L.)	008023801	172.510
GENET, EXTRACT (SPARTIUM JUNCEUM L.)	977091283	172.510
GENTIAN ROOT, EXTRACT (GENTIANA LUTEA L.)	977091294	172.510
GENTIAN, STEMLESS (GENTIANA ACAULIS L.)	977088417	172.510
GERANIOL	000106241	182.60
GERANIUM, EAST INDIAN, EXTRACT(CYMBOPOGON MARTINI STAPF.)	977091476	182.20
GERANIUM, EAST INDIAN, OIL(CYMBOPOGON MARTINI STAPF.)	008014195	182.20
GERANIUM EXTRACT (PELARGONIUM SPP.)	977091465	182.20
GERANIUM, OIL(PELARGONIUM SPP.)	008000462	182.20
GERANIUM (PELARGONIUM SPP.)	977001356	182.10
GERANIUM, ROSE, OIL (PELARGONIUM GRAVEOLENS L'HER.)	977143784	182.20
GERANYL ACETATE	000105873	182.60

MAINTERM	CAS	REGNUM
GERANYL ACETOACETATE	010032005	172.515
GERANYL ACETONE	003796701	172.515
GERANYL BENZOATE	000094484	172.515
GERANYL BUTYRATE	000106296	172.515
GERANYL FORMATE	000105862	172.515
GERANYL HEXANOATE	010032027	172.515
GERANYL ISOBUTYRATE	002345268	172.515
GERANYL ISOVALERATE	000109206	172.515
GERANYL PHENYLACETATE	000102227	172.515
GERANYL PROPIONATE	000105908	172.515
GERMANDER, CHAMAEDRYS, EXTRACT SOLID (TEUCRIUM CHAMAEDRYS L.)	977091523	172.510
GERMANDER, CHAMAEDRYS, EXTRACT (TEUCRIUM CHAMAEDRYS L.)	977091512	172.510
GERMANDER, CHAMAEDRYS (TEUCRIUM CHAMAEDRYS L.)	977081085	172.510
GERMANDER, GOLDEN (TEUCRIUM POLIUM L.)	977088440	172.510
GHATTI, GUM (ANOGEISSUS LATIFOLIA WALL.)	009000286	184.1333
GIBBERELLIC ACID & POTASSIUM GIBBERELLATE	977136814	172.725
GINGER, EXTRACT (ZINGIBER OFFICINALE ROSC.)	977023554	182.20
GINGER, OIL (ZINGIBER OFFICINALE ROSC.)	008007087	182.20
GINGER, OLEORESIN (ZINGIBER OFFICINALE ROSC.)	008002606	182.20
GINGER (ZINGIBER OFFICINALE ROSC.)	977001389	182.10
D-GLUCONIC ACID	000526954	
GLUCONO-DELTA LACTONE	000090802	133.129 131.144 184.1318

MAINTERM	CAS	REGNUM
GLUCOSE ISOMERASE FROM BACILLUS COAGULANS	977042639	
GLUCOSE ISOMERASE FROM IMMOBILIZED ARTHROBACTER GLOBIFORMIS	977090086	
GLUCOSE ISOMERASE FROM STREPTOMYCES OLIVACEUS	977078162	
GLUCOSE ISOMERASE FROM STREPTOMYCES OLIVOCHROMOGENES	977090075	
GLUCOSE ISOMERASE FROM STREPTOMYCES RUBIGINOSUS	977090064	
GLUCOSE OXIDASE CATALASE PREPARATION	009001358	
GLUCOSE OXIDASE FROM ASPERGILLUS NIGER	977031825	
GLUCOSE OXIDASE FROM PENICILLIUM NOTATUM	977090097	
GLUCOSE PENTAACETATE	000083874	172.515
GLUCOSIDASE FROM ASPERGILLUS FLAVUS	977091487	
GLUCOSIDASE FROM ASPERGILLUS NIGER	977091498	
GLUCOSIDASE FROM ASPERGILLUS ORYZAE	977091501	
L-GLUTAMIC ACID	000056860	182.1045 172.320
GLUTAMIC ACID HYDROCHLORIDE	000138158	172.320 182.1047
L-GLUTAMINE	000056859	172.320
GLUTARALDEHYDE	000111308	172.230 173.357 175.105 176.170 176.180 173.320
GLUTEN, GUM	977091567	139.150 139.140 139.110 139.115 139.117 139.120 139.125 139.135 139.155 139.160 139.165 139.180

MAINTERM	CAS	REGNUM
GLYCERIN	000056815	182.90 182.1320 169.175 169.176 169.177 169.178 169.180 169.181
GLYCERIN, SYNTHETIC	977091534	172.866
GLYCEROL LACTOPALMITATE	001338096	172.852
GLYCEROL TRIBUTYRATE	000060015	184.1903
GLYCERYL BEHENATE	077538193	184.1328
GLYCERYL 5-HYDROXYDECANOATE	026446311	
GLYCERYL 5-HYDROXYDODECANOATE	026446322	
GLYCERYL-LACTO ESTERS OF FATTY ACIDS	977051312	172.852
GLYCERYL LACTOOLEATE	030283160	172.852
GLYCERYL MONOOLEATE	025496724	184.1323
GLYCERYL MONOSTEARATE	031566311	139.150 184.1324 139.110 139.115 139.117 139.120 139.121 139.122 139.125 139.135 139.138 139.140 139.155 139.160 139.165 139.180
GLYCERYL TRIACETATE	000102761	184.1901
GLYCERYL TRIBENZOATE	000614335	
GLYCERYL TRIPROPANOATE	000139457	
GLYCERYL TRISTEARATE	000555431	
GLYCINE	000056406	172.812 172.320
GLYCOCHOLIC ACID	000475310	

MAINTERM	CAS	REGNUM
GLYCYRRHIZIN, AMMONIATED (GLYCYRRHIZA SPP.)	053956040	184.1408
GRAINS OF PARADISE (AFRAMOMUM MELEGUETA (ROSC.) K. SCHUM.)	977050875	182.10
GRAPE COLOR EXTRACT	977091578	73.169
GRAPE ESSENCE, NATURAL	977091545	
GRAPEFRUIT ESSENCE, NATURAL	977091556	182.20
GRAPEFRUIT, OIL (CITRUS PARADISI MACF.)	008016204	182.20
GRAPEFRUIT OIL, CONC.	977083058	182.20
GRAPE SKIN EXTRACT	977010528	73.170
GUAIAC GUM, EXTRACT (GUAIACUM SPP.)	977099152	172.510
GUAIAC GUM(GUAIACUM SPP.)	009000297	172.510
GUAIACOL	000090051	172.515
GUAIAC WOOD, EXTRACT (GUAIACUM SPP.)	977083525	172.510
GUAIAC WOOD, OIL (GUAIACUM SPP.)	008016237	172.510
GUAIACYL ACETATE	000613707	172.515
GUAIACYL PHENYLACETATE	004112894	172.515
GUAIENE	000088846	172.515
GUAIOL ACETATE	000134281	172.515
GUARANA, GUM (PAULLINIA CUPANA HBK)	888285254	172.510
GUAR, GUM (CYAMOPSIS TETRAGONOLOBUS (L.))	009000300	184.1339 150.161 150.141 133.124 133.133 133.134 133.162 133.178 133.179
GUAVA (PSIDIUM SPP.)	977050900	182.20
GUTTA HANG KANG (PALAQUIUM LEIOCARPUM BOERL. AND P.		

MAINTERM	CAS	REGNUM
OBLONGIFOLIUM BURCK.)	009007970	172.615
HAW BARK, BLACK, EXTRACT (VIBURNUM PRUNIFOLIUM L.)	977073850	172.510
HELIUM	007440597	184.1355
HEMLOCK NEEDLES AND TWIGS, OIL (TSUGA SPP.)	008008104	172.510
HEMLOCK (TSUGA SPP.)	977074626	172.510
2,4-HEPTADIENAL	004313035	
G&-HEPTALACTONE	000105215	172.515
HEPTANAL	000111717	172.515
HEPTANAL DIMETHYL ACETAL	010032050	172.515
HEPTANAL GLYCERYL ACETAL (MIXED 1,2 AND 1,3 ACETALS)	977043665	172.515
2,3-HEPTANEDIONE	000096048	172.515
HEPTANOIC ACID	000111148	173.315
2-HEPTANOL	000543497	
3-HEPTANOL	000589822	172.515
2-HEPTANONE	000110430	172.515
3-HEPTANONE	000106354	172.515
4-HEPTANONE	000123193	172.515
CIS-4-HEPTENAL	006728310	172.515
2-HEPTENAL	002463630	
4-HEPTENAL DIETHYL ACETAL	977044500	
2-HEPTEN-4-ONE	004643258	
3-HEPTEN-2-ONE	001119444	
TRANS-3-HEPTENYL ACETATE	001576778	
TRANS-3-HEPTENYL 2-METHYLPROPANOATE	977045638	
HEPTYL ACETATE	000112061	172.515

MAINTERM	CAS	REGNUM
HEPTYL ALCOHOL	000111706	172.515
HEPTYL BUTYRATE	005870939	172.515
HEPTYL CINNAMATE	010032083	172.515
3-HEPTYLDIHYDRO-5-METHYL-2(3H)-FUR-ANONE	040923646	
HEPTYL FORMATE	000112232	172.515
2-HEPTYLFURAN	003777717	
HEPTYL ISOBUTYRATE	002349135	172.515
HEPTYL OCTANOATE	004265978	172.515
HEPTYLPARABEN	001085127	172.145
HESPERIDIN	000520263	
1-HEXADECANOL	036653824	172.515 73.1001 73.1 172.864
OMEGA-6-HEXADECENLACTONE	007779502	172.515
TRANS, TRANS-2,4-HEXADIENAL	000142836	
D&-HEXALACTONE	000823223	
G&-HEXALACTONE	000695067	172.515
HEXANAL	000066251	172.515
HEXANE	000110543	172.560 173.270 172.340 175.105 175.320 176.200 177.1200
2,3-HEXANEDIONE	003848246	172.515
3,4-HEXANEDIONE	004437518	
1,6-HEXANEDITHIOL	001191431	
HEXANOIC ACID	000142621	173.315 172.515

MAINTERM	CAS	REGNUM
3-HEXANOL	000623370	
3-HEXANONE	000589388	
CIS-3-HEXENAL	006789806	
CIS-4-HEXENAL	004634893	
TRANS-3-HEXENAL	069112216	
2-HEXENAL	000505577	172.515
4-HEXENE-3-ONE	002497214	
TRANS-2-HEXENOIC ACID	013419697	
3-HEXENOIC ACID	004219243	
1-HEXEN-3-OL	004798441	
2-HEXEN-1-OL	002305217	172.515
4-HEXEN-1-OL	006126507	
CIS-3-HEXEN-1-OL	000928961	172.515
CIS-3-HEXEN-1-YL ACETATE	003681718	
TRANS-2-HEXEN-1-YL ACETATE	002497189	172.515
CIS-3-HEXENYL BENZOATE	025152856	
CIS-3-HEXENYL BUTYRATE	016491364	
CIS-3-HEXENYL FORMATE	033467731	
CIS-3-HEXENYL HEXANOATE	031501118	
CIS-3-HEXENYL CIS-3-HEXENOATE	061444380	
3-HEXENYL ISOVALERATE	010032118	172.515
CIS-3-HEXENYL LACTATE	061931815	
3-HEXENYL 2-METHYLBUTYRATE	010094414	172.515
3-HEXENYL PHENYLACETATE	042436077	172.515
CIS-3-HEXENYL PROPIONATE	033467742	
HEXYL ACETATE	000142927	172.515

MAINTERM	CAS	REGNUM
2-HEXYL-4-ACETOXYTETRAHYDROFURAN	010039391	172.515
HEXYL ALCOHOL	000111273	172.515 172.864
HEXYL BENZOATE	006789884	
N-HEXYL 2-BUTENOATE	019089920	
HEXYL BUTYRATE	002639636	172.515
A&-HEXYLCINNAMALDEHYDE	000101860	172.515
HEXYL FORMATE	000629334	172.515
HEXYL 2-FUROATE	039251860	
HEXYL HEXANOATE	006378650	172.515
HEXYL TRANS-2-HEXENOATE	033855571	
2-HEXYLIDENE CYCLOPENTANONE	017373896	172.515
HEXYL ISOBUTYRATE	002349077	
HEXYL ISOVALERATE	010032130	172.515
2-HEXYL-5 OR 6-KETO-1,4-DIOXANE	977043609	
HEXYL 2-METHYLBUTYRATE	010032152	172.515
HEXYL 2-METHYL-3(OR 4)-PENTENOATE	977101055	
HEXYL OCTANOATE	001117551	172.515
HEXYL PHENYLACETATE	005421170	172.515
HEXYL PROPIONATE	002445763	172.515
HICKORY BARK, EXTRACT (CARYA SPP.)	977023225	182.20
HICKORY SMOKE DIST.	074113749	
L-HISTIDINE	000071001	172.320
HOPS, EXTRACT (HUMULUS LUPULUS L.)	977070679	182.20
HOPS EXTRACT, MODIFIED	008016259	172.560
HOPS, EXTRACT SOLID (HUMULUS		

MAINTERM	CAS	REGNUM
LUPULUS L.)	977083252	182.20
HOPS, OIL (HUMULUS LUPULUS L.)	008007043	182.20
HOREHOUND EXTRACT (MARRUBIUM VULGARE L.)	977024853	182.20
HOREHOUND(MARRUBIUM VULGARE L.)	977001594	182.10
HOREHOUND SOLID, EXTRACT	977089410	182.20
HORSEMINT LEAVES, EXTRACT (MONARDA SPP.)	008006857	182.20
HORSERADISH (ARMORACIA LAPATHIFOLIA GILIB.)	977050944	182.10
HORSERADISH OIL	977089421	
HYACINTH, ABSOLUTE (HYACINTHUS ORIENTALIS L.)	977086466	172.510
HYACINTH FLOWERS (HYACINTHUS ORIENTALIS L.)	977047907	172.510
HYDRATROPIC ALDEHYDE PROPYLENE GLYCOL ACETAL	067634235	
HYDRAZINE	000302012	173.310
HYDROCHLORIC ACID	007647010	182.1057 172.560 172.892 160.105 133.129 160.185 131.144
HYDROGEN PEROXIDE	007722841	172.814 175.105 172.892 160.145 160.105 160.185 133.113 184.1366 178.1010
HYDROGEN SULFIDE	007783064	
HYDROQUINONE MONOETHYL ETHER	000622628	
2-HYDROXYACETOPHENONE	000118934	
4-HYDROXYBUTANOIC ACID LACTONE	000096480	
1-HYDROXY-2-BUTANONE	005077678	

MAINTERM	CAS	REGNUM
HYDROXYCITRONELLAL	000107755	172.515
HYDROXYCITRONELLAL DIETHYL ACETAL	007779944	172.515
HYDROXYCITRONELLAL DIMETHYL ACETAL	000141924	172.515
HYDROXYCITRONELLOL	000107744	172.515
2-HYDROXY-2-CYCLOHEXEN-1-ONE	010316662	
5-HYDROXY-2,4-DECADIENOIC ACID D&-LACTONE	027593233	
5-HYDROXY-2-DECENOIC ACID D&-LACTONE	051154962	
5-HYDROXY-7-DECENOIC ACID D&-LACTONE	025524952	
6-HYDROXYDIHYDROTHEASPIRANE	053398906	
4-HYDROXY-2,5-DIMETHYL-3(2H)-FURAN-ONE	003658773	
6-HYDROXY-3,7-DIMETHYLOCTANOIC ACID LACTONE	000499547	
1-HYDROXYETHYLIDENE-1,1-DIPHOSPHON-IC ACID	002809214	173.310
HYDROXYLATED LECITHIN	008029763	172.814 136.110 173.340 136.115 136.130 136.160 136.165 136.180
N-(4-HYDROXY-3-METHOXYBENZYL)-8-ME-THYL-6-NONENAMIDE	000404864	182.10 182.20
2-HYDROXY-4-METHYLBENZALDEHYDE	000698271	
2-HYDROXYMETHYL-6,6-DIMETHYLBICYCL-O(3.1.1)HEPT-2-ENYL FORMATE	072928520	
4-HYDROXYMETHYL-2,6-DI-TERTBUTYLPH-ENOL	000088266	172.150
4-HYDROXY-5-METHYL-3(2H)-FURANONE	019322271	
3-(HYDROXYMETHYL)-2-OCTANONE	059191785	
HYDROXYNONANOIC ACID, D&-LACTONE	003301948	

MAINTERM	CAS	REGNUM
5-HYDROXY-4-OCTANONE	000496775	172.515
3-HYDROXY-2-PENTANONE	003142663	
4-HYDROXY-3-PENTENOIC ACID LACTONE	000591128	
4-(P-HYDROXYPHENYL)-2-BUTANONE	005471512	172.515
L-HYDROXYPROLINE	000051354	
HYDROXYPROPYL CELLULOSE	009004642	172.870
HYDROXYPROPYL METHYLCELLULOSE	009004653	172.874
2-HYDROXY-3,5,5-TRIMETHYL-2-CYCLOH-EXENONE	004883607	
5-HYDROXYUNDECANOIC ACID LACTONE	000710043	
5-HYDROXY-8-UNDECENOIC ACID DELTA-LACTONE	068959284	
HYSSOP, EXTRACT (HYSSOPUS OFFICINALIS L.)	977083069	182.20
HYSSOP (HYSSOPUS OFFICINALIS L.)	977001630	182.10
HYSSOP, OIL (HYSSOPUS OFFICINALIS L.)	008006835	182.20
ICELAND MOSS (CETRARIA ISLANDICA ACH.)	977017632	172.510
IMMORTELLE, EXTRACT (HELICHRYSUM ANGUSTIFOLIUM DC.)	977030195	182.20
IMPERATORIA (PEUCEDANUM OSTRUTHIUM (L.) KOCH (IMPERATORIA OSTRUTHIUM L.))	977002326	172.510
INDOLE	000120729	172.515
INOSITOL	000087898	182.5370 184.1370
INVERTASE FROM SACCHAROMYCES CEREVISIAE	977122976	
INVERT SUGAR	008013170	PART 146 169.175 145.180 145.134 PART 145 145.181 169.176

MAINTERM	CAS	REGNUM
		169.177
		169.178
		169.180
		169.181
		184.1859
INVERT SUGAR SYRUP	977083547	182.90
ION EXCHANGE MEMBRANES	977089669	173.20
ION EXCHANGE RESIN	977017052	173.25
A&-IONOL	025312349	
B&-IONOL	022029761	
A&-IONONE	000127413	172.515
B&-IONONE	014901076	172.515
G&-IONONE	000079765	
IRON AMMONIUM CITRATE	001185575	573.560
		582.80
		172.430
IRON CAPRYLATE	006535202	181.25
IRON-CHOLINE CITRATE COMPLEX	001336807	573.580
		172.370
IRON CITRATE	002338058	
A&-IRONE	000079696	172.515
IRON, ELECTROLYTIC	977092957	184.1375
IRON LINOLEATE	007779637	181.25
IRON NAPHTHENATE	001338143	181.25
IRON OXIDE	001332372	177.2600
		177.1460
		582.80
		175.390
		175.380
		177.1350
		175.300
		182.99
		73.2250
IRON OXIDE, SYNTHETIC	977053385	73.1200
		73.200
IRON PEPTONATE	977009725	

MAINTERM	CAS	REGNUM
IRON POLYVINYLPYRROLIDONE	977125475	
IRON, REDUCED	007439896	182.5375
		582.80
		582.5375
		184.1375
IRON TALLATE	061788816	181.25
ISOAMYL ACETATE	000123922	172.515
ISOAMYL ACETOACETATE	002308181	172.515
ISOAMYL ALCOHOL	000123513	172.515
ISOAMYL BENZOATE	000094462	172.515
ISOAMYL BUTYRATE	000106274	172.515
ISOAMYL CINNAMATE	007779659	172.515
ISOAMYL FORMATE	000110452	172.515
ISOAMYL 4-(2-FURAN)BUTYRATE	007779660	172.515
ISOAMYL 3-(2-FURAN)PROPIONATE	007779671	172.515
ISOAMYL HEXANOATE	002198610	172.515
ISOAMYL ISOBUTYRATE	002050013	172.515
ISOAMYL ISOVALERATE	000659701	172.515
ISOAMYL LAURATE	006309519	172.515
ISOAMYL 2-METHYLBUTYRATE	027625350	172.515
ISOAMYL NONANOATE	007779706	172.515
ISOAMYL OCTANOATE	002035996	172.515
ISOAMYL PHENYLACETATE	000102192	172.515
ISOAMYL PROPIONATE	000105680	172.515
ISOAMYL PYRUVATE	007779728	172.515
ISOAMYL SALICYLATE	000087207	172.515
ISOBORNEOL	000124765	172.515
ISOBORNYL ACETATE	000125122	172.515

MAINTERM	CAS	REGNUM
ISOBORNYL FORMATE	001200675	172.515
ISOBORNYL ISOVALERATE	007779739	172.515
ISOBORNYL PROPIONATE	002756561	172.515
ISOBUTANE	000075285	184.1165
ISOBUTYL ACETATE	000110190	172.515
ISOBUTYL ACETOACETATE	007779751	172.515
ISOBUTYL ALCOHOL	000078831	172.515 73.1
ISOBUTYL ANGELATE	007779819	172.515
ISOBUTYL ANTHRANILATE	007779773	172.515
ISOBUTYL BENZOATE	000120503	172.515
ISOBUTYL 2-BUTENOATE	000589662	
ISOBUTYL BUTYRATE	000539902	172.515
ISOBUTYL CINNAMATE	000122678	172.515
ISOBUTYLENE-ISOPRENE COPOLYMER	009010859	172.615
ISOBUTYL FORMATE	000542552	172.515
ISOBUTYL 2-FURANPROPIONATE	000105011	172.515
ISOBUTYL HEPTANOATE	007779808	172.515
ISOBUTYL HEXANOATE	000105793	172.515
ISOBUTYL ISOBUTYRATE	000097858	172.515
2-ISOBUTYL-3-METHOXYPYRAZINE	024683009	
ISOBUTYL N-METHYLANTHRANILATE	065505240	
2-ISOBUTYL-3-METHYLPYRAZINE	013925069	
A&-ISOBUTYLPHENETHYL ALCOHOL	007779784	172.515
ISOBUTYL PHENYLACETATE	000102136	172.515
ISOBUTYL PROPIONATE	000540421	172.515

MAINTERM	CAS	REGNUM
ISOBUTYL SALICYLATE	000087194	172.515
2-ISOBUTYLTHIAZOLE	018640749	172.515
ISOBUTYRALDEHYDE	000078842	172.515
ISOBUTYRIC ACID	000079312	172.515
ISOCYCLOCITRAL	001335666	
ISOEUGENOL	000097541	172.515
ISOEUGENYL ACETATE	000093298	172.515
ISOEUGENYL BENZYL ETHER	000120116	172.515
ISOEUGENYL ETHYL ETHER	007784670	172.515
ISOEUGENYL FORMATE	007774961	172.515
ISOEUGENYL METHYL ETHER	000093163	172.515
ISOEUGENYL PHENYLACETATE	000120241	172.515
ISOJASMONE	011050627	172.515
DL-ISOLEUCINE	000443798	
L-ISOLEUCINE	000073325	172.320
DL-ISOMENTHONE	000491076	
A&-ISOMETHYLIONONE	000127515	172.515
ISOPENTYLAMINE	000107857	
ISOPHORONE	000078591	175.105
CIS-5-ISOPROPENYL-CIS-2-METHYLCYCL-OPENTAN-1-CARBOXALDEHYDE	055253286	
5-ISOPROPENYL-2-METHYL-2-VINYLTETR-AHYDROFURAN	013679862	
ISOPROPENYLPYRAZINE	038713416	
ISOPROPYL ACETATE	000108214	172.515
P-ISOPROPYLACETOPHENONE	000645136	172.515

MAINTERM	CAS	REGNUM
ISOPROPYL ALCOHOL	000067630	173.340 73.1 172.515 172.560 172.712 173.240 73.1001
ISOPROPYL BENZOATE	000939480	172.515
P-ISOPROPYLBENZYL ALCOHOL	000536607	172.515
ISOPROPYL BUTYRATE	000638119	172.515
ISOPROPYL CINNAMATE	007780065	172.515
ISOPROPYL CITRATE	039413053	182.6386 166.110 166.40 184.1386
ISOPROPYL FORMATE	000625558	172.515
ISOPROPYL HEXANOATE	002311468	172.515
ISOPROPYL ISOBUTYRATE	000617505	172.515
ISOPROPYL ISOVALERATE	032665239	172.515
ISOPROPYL 2-METHYLBUTYRATE	066576714	
2-ISOPROPYL-5-METHYL-2-HEXENAL	035158259	
2-ISOPROPYL-4-METHYLTHIAZOLE	015679137	
ISOPROPYL MYRISTATE	000110270	
ISOPROPYL PALMITATE	000142916	
2-ISOPROPYLPHENOL	000088697	
P-ISOPROPYLPHENYLACETALDEHYDE	004395920	172.515
ISOPROPYL PHENYLACETATE	004861852	172.515
3-(P-ISOPROPYLPHENYL)PROPIONALDEHY- DE	007775000	172.515
ISOPROPYL PROPIONATE	000637785	172.515
ISOPROPYL TIGLATE	001733251	
ISOPULEGOL	000089792	172.515

MAINTERM	CAS	REGNUM
ISOPULEGONE	029606799	172.515
ISOPULEGYL ACETATE	000089496	172.515
ISOQUINOLINE	000119653	172.515
ISOVALERIC ACID	000503742	172.515
IVA (ACHILLEA MOSCHATA JACQ.)	977091614	172.510
IVA, EXTRACT (ACHILLEA MOSCHATA JACQ.)	977091625	172.510
JAPAN WAX	008001396	182.70 175.105 175.350 176.170 73.1
JASMINE, ABSOLUTE (JASMINUM GRANDIFLORUM L.)	977146681	182.20
JASMINE, CONCRETE (JASMINUM GRANDIFLORUM L.)	977125384	182.20
JASMINE, OIL (JASMINUM GRANDIFLORUM L.)	008022966	182.20
JASMINE, SPIRITUS (JASMINUM GRANDIFLORUM L.)	977038791	182.20
JELUTONG (DYERA COSTULATA HOOK, F. AND D. LOWII HOOK, F.)	977011441	172.615
JUNIPER BERRIES (JUNIPERUS COMMUNIS L.)	977038804	182.20
JUNIPER, EXTRACT (JUNIPERUS COMMUNIS L.)	977009225	182.20
JUNIPER OIL (JUNIPERUS COMMUNIS L.)	008002684	182.20
KARAYA, GUM (STERCULIA URENS ROXB.)	009000366	133.133 184.1349 150.141 150.161 133.134 133.162 133.178 133.179
KELP	977001754	172.365 184.1120
2-KETO-4-BUTANETHIOL	034619120	
A&-KETOBUTYRIC ACID	000600180	

MAINTERM	CAS	REGNUM
KOLA NUT, EXTRACT (COLA ACUMINATA SCHOTT ET ENDL.)	977024831	182.20
LABDANUM, ABSOLUTE (CISTUS SPP.)	977046982	172.510
LABDANUM, OIL (CISTUS SPP.)	008016260	172.510
LABDANUM, OLEORESIN (CISTUS SPP.)	977092720	172.510
LACTALBUMIN	009013905	
LACTALBUMIN PHOSPHATE	977051005	
LACTASE FROM SACCHAROMYCES FRAGILIS	977090100	
LACTASE FROM SACCHAROMYCES (KLUYVEROMYCES) LACTIS	977090111	184.1388
LACTIC ACID	000050215	150.161 PART 133 150.141 184.1061 131.144 172.814
LACTOSE	000063423	133.179 133.178 133.124 169.179 169.182
LACTOSE, HYDROLYZED	977126934	133.179 133.178 133.124
LACTYLATED FATTY ACID ESTERS OF GLYCEROL AND PROPYLENE GLYCOL	977050660	172.850
LACTYLIC ESTERS OF FATTY ACIDS	977050671	172.848
LANOLIN	008020846	172.615
LARD	061789999	182.70
LARD OIL	008016282	182.70
LAUREL BERRIES (LAURUS NOBILIS L.)	977051016	182.20
LAURIC ACID	000143077	173.340 172.210 172.860
LAURIC ALDEHYDE	000112549	172.515
LAURYL ACETATE	000112663	172.515

MAINTERM	CAS	REGNUM
LAURYL ALCOHOL	000112538	172.864 172.515
LAVANDIN, OIL	008022159	182.20
LAVENDER, ABSOLUTE (LAVANDULA OFFICINALIS CHAIX)	977126263	182.20
LAVENDER, CONCRETE (LAVANDULA OFFICINALIS CHAIX)	977089329	182.20
LAVENDER (LAVANDULA OFFICINALIS CHAIX)	977001823	182.10
LAVENDER, OIL (LAVANDULA OFFICINALIS CHAIX)	008000280	182.20
LAVENDER, SPIKE (LAVANDULA LATIFOLIA BILL.)	977051050	182.20
LAVENDER, SPIKE, OIL (LAVANDULA SPP.)	008016782	182.20
LECHE CASPI (COUMA MACROCARPA BARB. RODR.)	977011452	172.615
LECHE DE VACA (BROSIMUM UTILE (H.B.K.) PITTIER, AND POULSENIA SPP.)	977011463	172.615
LECITHIN	008002435	184.1400 166.110 169.115 136.110 133.169 163.130 163.123 166.40 133.179 133.173 169.140 169.150 136.115 136.130 136.160 136.165 136.180 163.135 163.140 163.145 163.155 163.150 163.153
LECITHIN, BENZOYL PEROXIDE MODIFIED	977092753	184.1400
LECITHIN, HYDROGEN PEROXIDE MODIFIED	977092764	
LECITHIN (VEGETABLE)	977092242	184.1400

MAINTERM	CAS	REGNUM
LEEK OIL	977089432	
LEMON ESSENCE	977091761	
LEMON, EXTRACT (CITRUS LIMON (L.) BURM. F.)	008028384	182.20
LEMON-GRASS, OIL (CYMBOPOGON CITRATUS DC. AND CYMBOPOGON FLEXUOSUS STAPF)	008007021	182.20
LEMON, OIL (CITRUS LIMON (L.) BURM. F.)	008008568	182.20 161.190 146.114 146.121 146.120 146.126
LEMON, OIL, TERPENELESS (CITRUS LIMON (L.) BURM. F.)	068648395	
LEMON PEEL EXTRACT	977091772	182.20
LEMON PEEL GRANULES	977001834	172.510
LEMON-VERBENA (LIPPIA CITRIODORA HBK.)	977047963	172.510
L-LEUCINE	000061905	172.320
LEVULINIC ACID	000123762	172.515
LEVULOSE	000057487	
LICORICE EXTRACT (GLYCYRRHIZA SPP.)	068916916	184.1408
LICORICE EXTRACT POWDER (GLYCYRRHIZA SPP.)	977070624	184.1408
LICORICE (GLYCYRRHIZA SPP.)	977004311	184.1408
LIGNIN	009005532	
LIGNOSULFONIC ACID	008062155	173.310
LIME JUICE, DEHYDRATED	977091783	182.20
LIME OIL, DISTILLED	008008262	182.20
LIME OIL, EXPRESSED	977059805	182.20
LIME, OIL, TERPENELESS (CITRUS AURANTIFOLIA (CHRISTMAN) SWINGLE)	068916847	

MAINTERM	CAS	REGNUM
D-LIMONENE	005989275	182.60
DL-LIMONENE	007705148	182.60
L-LIMONENE	005989548	182.60
LINALOE WOOD, OIL (BURSERA DELPECHIANA POISS. AND OTHER BURSERA SPP.)	977051129	172.510
LINALOOL	000078706	182.60
LINALOOL OXIDE	001365191	172.515
LINALYL ACETATE	000115957	182.60
LINALYL ANTHRANILATE	007149260	172.515
LINALYL BENZOATE	000126647	172.515
LINALYL BUTYRATE	000078364	172.515
LINALYL CINNAMATE	000078375	172.515
LINALYL FORMATE	000115991	172.515
LINALYL HEXANOATE	007779239	172.515
LINALYL ISOBUTYRATE	000078353	172.515
LINALYL ISOVALERATE	001118270	172.515
LINALYL OCTANOATE	010024643	172.515
LINALYL PHENYLACETATE	007143693	
LINALYL PROPIONATE	000144398	172.515
LINDEN FLOWERS, EXTRACT (TILIA SPP.)	977060573	182.20
LINDEN FLOWERS (TILIA GLABRA VENT.)	977009770	182.10
LINDEN LEAVES (TILLIA SPP.)	977073429	172.510
LINOLEIC ACID	000060333	182.5065 184.1065
LIPASE	009001621	
LIPASE FROM ANIMAL TISSUE	977033785	

MAINTERM	CAS	REGNUM
LIPASE FROM ASPERGILLUS NIGER	977031563	
LIPASE FROM ASPERGILLUS ORYZAE	977031687	
LOVAGE, EXTRACT (LEVISTICUM OFFICINALE KOCH)	977091636	172.510
LOVAGE (LEVISTICUM OFFICINALE KOCH)	977048477	172.510
LOVAGE, OIL (LEVISTICUM OFFICINALE KOCH)	008016317	172.510
LUNGMOSS (STICTA PULMONACEA ACH.)	977022857	172.510
LUPULIN (HUMULUS LUPULUS L.)	977051130	182.20
L-LYSINE	000056871	172.320
MACE (MYRISTICA FRAGRANS HOUTT.)	977051141	182.10
MACE, OIL (MYRISTICA FRAGRANS HOUTT.)	977051152	182.20
MACE, OLEORESIN (MYRISTICA FRAGRANS HOUTT.)	977010608	182.20
MAGNESIUM CAPRATE	042966303	172.863
MAGNESIUM CAPRYLATE	003386570	172.863
MAGNESIUM CARBONATE	039409820	184.1425 163.110 133.102 137.105 133.106 133.111 133.141 133.165 133.181 133.183 133.195 137.155 137.165 137.160 137.170 137.175 137.180 137.185
MAGNESIUM CHLORIDE	007786303	184.1426 172.560
MAGNESIUM CYCLAMATE--PROHIBITED	007757859	189.135
MAGNESIUM FUMARATE	006880235	172.350
MAGNESIUM GLUCONATE	003632915	

MAINTERM	CAS	REGNUM
MAGNESIUM GLYCEROPHOSPHATE	000927208	
MAGNESIUM HYDROXIDE	001309428	184.1428
MAGNESIUM LAURATE	004040486	172.863
MAGNESIUM MYRISTATE	004086708	172.863
MAGNESIUM OLEATE	001555539	172.863
MAGNESIUM OXIDE	001309484	184.1431 182.5431 163.110
MAGNESIUM PALMITATE	002601981	172.863
MAGNESIUM PHOSPHATE, DIBASIC	007782754	182.5434 184.1434
MAGNESIUM PHOSPHATE, TRIBASIC	007757871	182.5434 184.1434
MAGNESIUM SALTS OF FATTY ACIDS	977093381	172.863
MAGNESIUM SILICATE	001343880	182.2437 169.179 169.182
MAGNESIUM STEARATE	000557040	173.340 184.1440 172.863
MAGNESIUM SULFATE	010034998	182.5443 184.1443
MAIDENHAIR FERN (ADIANTUM CAPILLUS-VENERIS L.)	977070306	172.510
L-MALIC ACID	000097676	150.141 169.140 169.115 150.161 146.113 169.150 184.1069
MALIC ACID	000617481	184.1069 131.144
MALT DIASTASE	009000924	
MALTODEXTRIN	009050366	184.1444
MALTOL	000118718	172.515
MALTOSE	000069794	133.124 133.178

MAINTERM	CAS	REGNUM
MALT SYRUP	008002480	184.1445 133.178 73.85
MALTYL ISOBUTYRATE	065416140	
MANDARIN, OIL (CITRUS RETICULATA BLANCO)	008008319	182.20
MANGANESE CHLORIDE	007773015	182.5446 582.80 184.1446
MANGANESE CITRATE	010024665	182.5449 184.1449
MANGANESE GLUCONATE	006485398	182.5452 582.80 184.1452
MANGANESE GLYCEROPHOSPHATE	001320463	182.5455 182.8455
MANGANESE HYPOPHOSPHITE	010043842	582.5458 182.8458
MANGANESE SULFATE	007785877	182.5461 582.80 184.1461
MANGANOUS OXIDE	001344430	182.5464 582.80
MANNITOL	000069658	180.25
MARIGOLD, POT (CALENDULA OFFICINALIS L.)	977001936	182.10
MARJORAM, OLEORESIN (MARJORANA HORTENSIS MOENCH (ORIGANUM MAJORANA L.))	977038859	182.20
MARJORAM, POT (MAJORANA ONITES (L.) BENTH. (ORIGANUM VULGARE L.))	977051221	182.10
MARJORAM SEED (MAJORANA HORTENSIS MOENCH (ORIGANUM MAJORANA L.))	977038860	182.10
MARJORAM, SWEET (MAJORANA HORTENSIS MOENCH (ORIGANUM MAJORANA L.))	977051232	182.10
MARJORAM, SWEET, OIL (MAJORANA HORTENSIS MOENCH (ORIGANUM MAJORANA L.))	008015018	182.20
MASSARANDUBA BALATA (MANILKARA HUBERI (DUCKE) CHEVALIER)	977011474	172.615

MAINTERM	CAS	REGNUM
MASSARANDUBA BALATA, SOLVENT-FREE RESIN EXTRACT	977044533	172.615
MASSARANDUBA CHOCOLATE (MANILKARA SOLIMOESENSIS GILLY)	008029832	172.615
MASSOIA BARK	977103802	
MASTIC GUM	061789922	
MATE, ABSOLUTE (ILEX PARAGUARIENSIS ST. HIL.)	977146670	182.20
MATE, LEAVES	977051243	
MENADIOL SODIUM DIPHOSPHATE	000131135	
MENHADEN OIL, HYDROGENATED	093572533	184.1472
MENHADEN OIL, PARTIALLY HYDROGENATED	977131057	184.1472
MENTHADIENOL	003269907	172.515
P-MENTHA-1,8-DIEN-7-OL	000536594	172.515
MENTHADIENYL ACETATE	015111974	172.515
P-MENTHAN-2-ONE	000499707	
P-MENTHA-8-THIOL-3-ONE	038462225	
P-MENTH-1-ENE-9-AL	029548149	
1-P-MENTHENE-8-THIOL	071159905	
P-MENTH-1-EN-3-OL	000491043	
P-MENTH-3-EN-1-OL	000586823	172.515
1-P-MENTHEN-9-YL ACETATE	028839136	172.515
MENTHOL	000089781	172.515 182.20
MENTHONE	000089805	172.515
MENTHYL ACETATE	000089485	172.515 182.20
MENTHYL ISOVALERATE	016409464	172.515

MAINTERM	CAS	REGNUM
L-MENTHYL LACTATE	059259380	
2-MERCAPTO-3-BUTANOL	054812861	
3-MERCAPTO-2-BUTANONE	040789988	
2-MERCAPTOMETHYLPYRAZINE	059021022	
3-MERCAPTO-2-PENTANONE	067633970	
2,3 OR 10-MERCAPTOPINANE	977136052	
2-MERCAPTOPROPIONIC ACID	000079425	
METHACRYLIC ACID-DIVINYLBENZENE COPOLYMER	050602216	173.25 172.775
DL-METHIONINE	000059518	172.320
L-METHIONINE	000063683	172.320
O-METHOXYBENZALDEHYDE	000135024	172.515
P-METHOXYBENZALDEHYDE	000123115	172.515
O-METHOXYCINNAMALDEHYDE	001504741	172.515
P-METHOXYCINNAMALDEHYDE	001963366	
2-METHOXY-(3 OR 5 OR 6)-ISOPROPYLPYRAZINE	977044453	
P-METHOXY-A&-METHYLCINNAMALDEHYDE	065405676	
2-METHOXY-4-METHYLPHENOL	000093516	172.515
2-METHOXY-3-(1-METHYLPROPYL)PYRAZINE	024168705	
(2 OR 5 OR 6)-METHOXY-3-METHYLPYRAZINE (MIXTURE OF ISOMERS)	977044497	
4-(P-METHOXYPHENYL)-2-BUTANONE	000104201	172.515
1-(4-METHOXYPHENYL)-4-METHYL-1-PENTEN-3-ONE	000103139	172.515
1-(P-METHOXYPHENYL)-1-PENTEN-3-ONE	000104278	172.515
1-(P-METHOXYPHENYL)-2-PROPANONE	000122849	172.515

MAINTERM	CAS	REGNUM
2-METHOXY-4-PROPYLPHENOL	002785877	
METHOXYPYRAZINE	003149288	
2-METHOXY-4-VINYLPHENOL	007786610	172.515
METHYL ACETATE	000079209	172.515
4'-METHYLACETOPHENONE	000122009	172.515
METHYL 1-ACETOXYCYCLOHEXYL KETONE	052789738	
1-METHYL-2-ACETYLPYRROLE	000932161	
METHYL ACRYLATE	000096333	
METHYL ACRYLATE-DIVINYLBENZENE, COMPLETELY HYDROLYZED, COPOLYMER	977083070	173.25
METHYL ACRYLATE-DVB-ACRYLONITRILE, COMPLETELY HYDROLYZED, TERPOLYMER	977092708	173.25
METHYL ACRYLATE-DVB(2%), COPOLYMER, AMINOLYZED WITH DMAPA	977083092	173.25
METHYL ACRYLATE-DVB(3.5%), COPOLYMER, AMINOLYZED WITH DMAPA	977083105	173.25
METHYL ACRYLATE-DVB-(DEG-DIVINYL ETHER), AMINOLYZED, TERPOLYMER	977083116	173.25
METHYL ALCOHOL	000067561	173.250 172.560 173.385
2-METHYLALLYL BUTYRATE	007149293	172.515
METHYL ANISATE	000121982	172.515
O-METHYLANISOLE	000578585	172.515
P-METHYLANISOLE	000104938	172.515
METHYL ANTHRANILATE	000134203	182.60
METHYLATED SILICA	977047203	
METHYL BENZOATE	000093583	172.515
A&-METHYLBENZYL ACETATE	000093925	172.515
METHYLBENZYL ACETATE (MIXED O-, M-, P-)	029759113	172.515

MAINTERM	CAS	REGNUM
A&-METHYLBENZYL ALCOHOL	000098851	172.515
A&-METHYLBENZYL BUTYRATE	003460444	172.515
METHYL BENZYL DISULFIDE	000699105	
A&-METHYLBENZYL FORMATE	007775384	172.515
A&-METHYLBENZYL ISOBUTYRATE	007775395	172.515
A&-METHYLBENZYL PROPIONATE	000120456	172.515
4-METHYLBIPHENYL	000644086	
2-METHYL-1-BUTANETHIOL	001878188	
3-METHYL-2-BUTANETHIOL	002084186	
3-METHYL-2-BUTANOL	000598754	
2-METHYL-2-BUTENAL	000497030	
3-METHYL-2-BUTENAL	000107868	
TRANS-2-METHYL-2-BUTENOIC ACID	000080591	
2-METHYL-3-BUTEN-2-OL	000115184	172.515
3-METHYL-2-BUTEN-1-OL	000556821	
3-METHYL-3-BUTEN-2-ONE	000814788	
2-METHYLBUTYL ACETATE	000624419	
2-METHYLBUTYL ISOVALERATE	002445774	172.515
2-METHYLBUTYL 2-METHYLBUTYRATE	002445785	
METHYL P-TERT-BUTYLPHENYLACETATE	003549233	172.515
2-METHYLBUTYRALDEHYDE	000096173	172.515
3-METHYLBUTYRALDEHYDE	000590863	172.515
METHYL BUTYRATE	000623427	172.515
2-METHYLBUTYRIC ACID	000116530	172.515
A&-METHYLCINNAMALDEHYDE	000101393	172.515

MAINTERM	CAS	REGNUM
P-METHYLCINNAMALDEHYDE	001504752	172.515
METHYL CINNAMATE	000103264	172.515
6-METHYLCOUMARIN	000092488	
3-METHYLCROTONIC ACID	000541479	
1-METHYL-1,3-CYCLOHEXADIENE	001489561	172.515
1-METHYL-2,3-CYCLOHEXADIONE	003008433	
METHYL CYCLOHEXANECARBOXYLATE	004630824	
3-METHYL-2-CYCLOHEXEN-1-ONE	001193186	
3-METHYL-1-CYCLOPENTADECANONE	000956821	
METHYLCYCLOPENTENOLONE	000765708	172.515
1-METHYL-1-CYCLOPENTEN-3-ONE	002758181	
METHYL 2-DECENOATE	002482395	
5H-5-METHYL-6,7-DIHYDROCYCLOPENTA(-B)PYRAZINE	023747480	
METHYL DIHYDROJASMONATE	024851987	
4-METHYL-2,6-DIMETHOXYPHENOL	006638057	
METHYL 3,7-DIMETHYL-6-OCTENOATE	002270602	
METHYL DISULFIDE	000624920	172.515
2-METHYL-1,3-DITHIOLANE	005616513	
METHYLENE CHLORIDE	000075092	172.560 173.255 73.1 175.105 177.1580 182.99
4-(3,4-METHYLENEDIOXYPHENYL)-2-BUT-ANONE	055418525	
METHYL ESTERS OF FATTY ACIDS (EDIBLE)	977050682	172.225
2-METHYL-(3 OR 5 OR 6)-ETHOXYPYRAZINE	977044464	

MAINTERM	CAS	REGNUM
2-METHYL-3-FURANTHIOL	028588741	
METHYL FURFURACRYLATE	000623187	
5-METHYLFURFURAL	000620020	
METHYL FURFURYL DISULFIDE	057500002	
2-METHYL-3 OR 5 OR 6-(FURFURYLTHIO)PYRAZINE (MIXTURE OF ISOMERS)	059035982	
METHYL 2-FUROATE	000611132	
2-METHYL-3-(2-FURYL)ACROLEIN	000874668	
3-(5-METHYL-2-FURYL)-BUTANAL	031704800	
3-((2-METHYL-3-FURYL)THIO)-4-HEPTA-NONE	061295418	
4-((2-METHYL-3-FURYL)THIO)-5-NONAN-ONE	061295509	
METHYL GLUCOSIDE-COCONUT OIL ESTER	008028431	172.816
6-METHYL-3,5-HEPTADIEN-2-ONE	001604280	172.515
METHYL HEPTANOATE	000106730	172.515
2-METHYLHEPTANOIC ACID	001188029	172.515
6-METHYL-5-HEPTEN-2-OL	001569604	172.515
6-METHYL-5-HEPTEN-2-ONE	000110930	172.515
5-METHYL-2-HEPTEN-4-ONE	081925817	
5-METHYL-2,3-HEXANEDIONE	013706860	
METHYL HEXANOATE	000106707	172.515
2-METHYLHEXANOIC ACID	004536236	
5-METHYLHEXANOIC ACID	000628466	
METHYL 2-HEXENOATE	002396772	172.515
METHYL 3-HEXENOATE	002396783	
5-METHYL-3-HEXEN-2-ONE	005166530	

MAINTERM	CAS	REGNUM
5-METHYL-5-HEXEN-2-ONE	003240093	
METHYL P-HYDROXYBENZOATE	000099763	184.1490 172.515 150.161 150.141
METHYL 3-HYDROXYHEXANOATE	021188589	
METHYL 2-HYDROXY-4-METHYLPENTANOATE	040348729	
A&-METHYL-B&-HYDROXYPROPYL A&-METHYL-B&-MERCAPTOPROPYL SULFIDE	054957027	
METHYL-A&-IONONE	000127424	172.515
METHYL-B&-IONONE	000127435	172.515
METHYL-D&-IONONE	007784987	172.515
METHYL ISOBUTYRATE	000547637	172.515
2-METHYL-3-(P-ISOPROPYLPHENYL) PROPIONALDEHYDE	000103957	172.515
2-METHYL-5-ISOPROPYLPYRAZINE	013925058	
METHYL ISOVALERATE	000556241	172.515
METHYL JASMONATE	001211296	
METHYL LAURATE	000111820	172.515
METHYL LINOLEATE (48%) METHYL LINOLENATE (52%) MIXTURE	977136803	
METHYL MERCAPTAN	000074931	172.515
METHYL O-METHOXYBENZOATE	000606451	172.515
1-METHYL-3-METHOXY-4-ISOPROPYLBENZ- ENE	001076568	
2-METHYL-5-METHOXYTHIAZOLE	038205640	
METHYL N-METHYLANTHRANILATE	000085916	172.515
METHYL 2-METHYLBUTYRATE	000868575	172.515
METHYL 2-METHYL-3-FURYL DISULFIDE	065505171	
METHYL 2-METHYLPENTANOATE	002177777	

MAINTERM	CAS	REGNUM
METHYL 4-(METHYLTHIO)BUTYRATE	053053513	
METHYL 2-METHYLTHIOBUTYRATE	051534668	
2-METHYL-5-(METHYLTHIO)FURAN	013678596	
METHYL 3-METHYLTHIOPROPIONATE	013532188	172.515
METHYL 4-METHYLVALERATE	002412808	172.515
METHYL MYRISTATE	000124107	172.515
1-METHYLNAPHTHALENE	000090120	
METHYL B&-NAPHTHYL KETONE	000093083	172.515
METHYL NICOTINATE	000093607	
METHYL NONANOATE	001731846	172.515
4-METHYLNONANOIC ACID	045019281	
METHYL 2-NONENOATE	000111795	172.515
METHYL 3-NONENOATE	013481873	
METHYL 2-NONYNOATE	000111808	172.515
2-METHYLOCTANAL	007786290	172.515
METHYL OCTANOATE	000111115	172.515
4-METHYLOCTANOIC ACID	054947749	
2-METHYL-2-OCTENAL	073757274	
METHYL CIS-4-OCTENOATE	021063718	
METHYL 2-OCTENOATE	002396852	
METHYL TRANS-2-OCTENOATE	007367819	
METHYL 2-OCTYNOATE	000111126	172.515
2-METHYL-(3 OR 5 OR 6)-(METHYLTHIO)PYRAZINE (MIXTURE OF ISOMERS)	067952652	
METHYL 2-OXO-3-METHYLPENTANOATE	003682426	
2-METHYLPENTANAL	000123159	

MAINTERM	CAS	REGNUM
4-METHYL-2,3-PENTANEDIONE	007493585	172.515
3-METHYLPENTANOIC ACID	000105431	
4-METHYLPENTANOIC ACID	000646071	
3-METHYL-1-PENTANOL	000589355	
4-METHYL-2-PENTANONE	000108101	172.515
2-METHYL-2-PENTENAL	000623369	
4-METHYL-2-PENTENAL	005362561	
2-METHYL-2-PENTENOIC ACID	003142721	
2-METHYL-3-PENTENOIC ACID	037674638	
2-METHYL-4-PENTENOIC ACID	001575742	
4-METHYL-3-PENTEN-2-ONE	000141797	
4-METHYL-4-PENTEN-2-ONE	003744023	
3-METHYL-2-(2-PENTENYL)-2-CYCLOPEN-TEN-1-ONE	000488108	172.515
4-METHYL-2-PENTYL-1,3-DIOXOLAN	001599491	
B&-METHYLPHENETHYL ALCOHOL	001123859	172.515
A&-METHYLPHENETHYL BUTYRATE	068922112	
METHYL PHENETHYL ETHER	003558609	
METHYL PHENYLACETATE	000101417	172.515
2-METHYL-4-PHENYL-2-BUTANOL	000103059	
3-METHYL-4-PHENYL-3-BUTENE-2-ONE	001901264	172.515
2-METHYL-4-PHENYL-2-BUTYL ACETATE	000103071	172.515
2-METHYL-4-PHENYL-2-BUTYL ISOBUTYRATE	010031717	172.515
2-METHYL-4-PHENYLBUTYRALDEHYDE	040654828	
3-METHYL-2-PHENYLBUTYRALDEHYDE	002439443	172.515
METHYL 4-PHENYLBUTYRATE	002046175	172.515

MAINTERM	CAS	REGNUM
5-METHYL-2-PHENYL-2-HEXENAL	021834924	
4-METHYL-1-PHENYL-2-PENTANONE	005349622	172.515
4-METHYL-2-PHENYL-2-PENTENAL	026643914	
METHYL 3-PHENYLPROPIONATE	000103253	172.515
METHYLPOLYSILICONE	009004733	
METHYL PROPENYL DISULFIDE	005905475	
METHYL PROPIONATE	000554121	172.515
3-METHYL-5-PROPYL-2-CYCLOHEXEN-1-O-NE	003720169	172.515
METHYL PROPYL DISULFIDE	U02179604	
2-METHYLPROPYL-3-METHYLBUTYRATE	000589593	
2-METHYL-4-PROPYL-1,3-OXATHIANE	067715804	
2-(2-METHYLPROPYL)PYRIDINE	006304241	
3-(2-METHYLPROPYL)PYRIDINE	014159616	
2-(1-METHYLPROPYL)THIAZOLE	018277275	
METHYL PROPYL TRISULFIDE	017619362	
2-METHYLPYRAZINE	000109080	
METHYL 2-PYRROLYL KETONE	001072839	
6-METHYLQUINOLINE	000091623	
4-METHYLQUINOLINE	000491350	
5-METHYLQUINOXALINE	013708128	
METHYL SALICYLATE	000119368	
METHYL SORBATE	000689894	
METHYL SULFIDE	000075183	172.515
2-METHYLTETRAHYDROFURAN-3-ONE	003188009	
7-METHYL-4,4A,5,6-TETRAHYDRO-2(3H)-		

MAINTERM	CAS	REGNUM
-NAPHTHALENONE	034545885	
2-METHYLTETRAHYDROTHIOPHEN-3-ONE	013679851	
4-METHYLTHIAZOLE	000693958	
4-METHYL-5-THIAZOLEETHANOL	000137008	
4-METHYL-5-THIAZOLEETHANOL ACETATE	000656531	
2-METHYLTHIOACETALDEHYDE	023328623	
2-METHYL-3-THIOACETOXY-4,5-DIHYDRO-FURAN	026486146	
3-(METHYLTHIO)BUTANAL	016630527	
4-(METHYLTHIO)BUTANAL	042919642	
4-(METHYLTHIO)BUTANOL	020582858	
1-(METHYLTHIO)-2-BUTANONE	013678585	
4-(METHYLTHIO)-2-BUTANONE	034047397	
METHYL THIOBUTYRATE	002432511	
METHYL 2-THIOFUROATE	013679613	
3-(METHYLTHIO)-1-HEXANOL	051755669	
2-((METHYLTHIO)METHYL)-2-BUTENAL	040878726	
4-(METHYLTHIO)-4-METHYL-2-PENTANONE	023550405	
2-(METHYLTHIOMETHYL)-3-PHENYLPROPE-NAL	065887083	
5-METHYL-2-THIOPHENECARBOXALDEHYDE	013679704	
O-(METHYLTHIO)PHENOL	001073296	
3-(METHYLTHIO)PROPANOL	000505102	
3-(METHYLTHIO)PROPIONALDEHYDE	003268493	172.515
3-METHYLTHIOPROPYL ISOTHIOCYANATE	000505793	
2-METHYL-3-TOLYLPROPIONALDEHYDE (MIXED O-, M-, P-)	977044511	172.515
3-METHYL-1,2,4-TRITHIANE	043040013	

MAINTERM	CAS	REGNUM
2-METHYLUNDECANAL	000110418	172.515
METHYL 9-UNDECENOATE	005760509	172.515
METHYL 2-UNDECYNOATE	010522186	172.515
METHYL VALERATE	000624248	172.515
2-METHYLVALERIC ACID	000097610	172.515
2-METHYL-5-VINYLPYRAZINE	013925081	
4-METHYL-5-VINYLTHIAZOLE	001759280	
MICROPARTICULATED PROTEIN PRODUCT	977132834	184.1498
MILK CLOTTING ENZYME FROM BACILLUS CEREUS (FRANKLAND AND FRANKLAND)	977017745	173.150(A)(2)
MILK CLOTTING ENZYME FROM ENDOTHIA PARASITICA	977017734	173.150(A)(1)
MILK CLOTTING ENZYME FROM MUCOR MIEHEI COONEY ET EMERSON	977017767	173.150(A)(4)
MILK CLOTTING ENZYME FROM MUCOR PUSILLUS L.	977017756	173.150(A)(3)
MILK POWDER, WHOLE, ENZYME-MODIFIED	977053443	
MIMOSA, ABSOLUTE (ACACIA DECURRENS WILLD. VAR. DEALBATA)	008031036	172.510
MINERAL OIL, WHITE	008012951	172.878 173.340 175.230 573.680 178.3570 178.3620
MOLASSES, CONCENTRATE	977083127	182.20
MOLASSES, EXTRACT(SACCHARUM OFFICINARUM L.)	977091603	182.20
MOLASSES (SACCHARUM OFFICINARUM L.)	977001992	182.20
MOLECULAR SIEVE RESINS	068954245	173.40
MONOAMMONIUM GLUTAMATE	007558636	182.1500
MONO- AND DIGLYCERIDES	067254733	163.123 172.755 184.1505

MAINTERM	CAS	REGNUM
		166.40
		163.130
		136.110
		136.115
		136.130
		136.160
		136.180
		163.135
		163.140
		163.145
		163.155
		163.150
		163.153
MONO- AND DIGLYCERIDES, ACETIC ACID ESTERS AND SODIUM AND CALCIUM SALTS	977093267	
MONO- AND DIGLYCERIDES, ACETYLTARTARIC ACID ESTERS AND SODIUM AND CALCIUM SALTS	977093278	
MONO- AND DIGLYCERIDES, CITRIC ACID ESTERS AND SODIUM AND CALCIUM SALTS	977093289	
MONO- AND DIGLYCERIDES, DIACETYLTARTARIC ACID ESTERS	977051298	136.110
		184.1101
		136.115
		136.130
		136.160
		136.165
		136.180
MONO- AND DIGLYCERIDES, ETHOXYLATED	061163335	172.834
MONO- AND DIGLYCERIDES, LACTIC ACID ESTERS AND SODIUM AND CALCIUM SALTS	977093290	
MONO- AND DIGLYCERIDES, MONOSODIUM PHOSPHATE DERIVATIVES	977051323	184.1521
MONO- AND DIGLYCERIDES, SODIUM SULFOACETATE DERIVATIVES	977052473	166.110
		166.40
MONOCHLOROACETIC ACID--PROHIBITED	000079118	189.155
MONOETHANOL AMINE	000141435	173.315(A)(3)
		173.315
MONOGLYCERIDE CITRATE	036291324	172.832
		176.170
		176.180
MONOGLYCERIDES, ACETYLATED	977051345	172.828
		175.230
MONOISOPROPYL CITRATE	001321579	182.6511
MONOPOTASSIUM GLUTAMATE	019473495	172.320
		182.1516

MAINTERM	CAS	REGNUM
MONOSODIUM GLUTAMATE	000142472	172.320 182.1 169.150 169.140 169.115 161.190 158.170 155.200 155.170 155.120 155.130 145.131
MORPHOLINE	000110918	173.310
MORPHOLINE, FATTY ACID SALTS	977034722	172.235
MOUNTAIN MAPLE (ACER SPICATUM LAM.)	977048488	172.510
MOUNTAIN MAPLE BARK (ACER SPICATUM LAM.)	977089647	172.510
MOUNTAIN MAPLE, EXTRACT SOLID (ACER SPICATUM LAM.)	977089658	172.510
MULLEIN FLOWERS (VERBASCUM SPP.)	977048466	172.510
MUSK AMBRETTE	000123693	
MUSK, KETONE	000081141	
MUSK TONQUIN (MOSCHUS MOSCHIFERUS L.)	008001045	182.50
MUSTARD, BROWN (BRASSICA SPP.)	977051389	182.10
MUSTARD, BROWN, EXTRACT (BRASSICA SPP.)	977091794	182.20
MUSTARD FLOUR	977071796	
MUSTARD OIL	008007407	182.20
MUSTARD, ORIENTAL	977088951	
MUSTARD, YELLOW (BRASSICA SPP.)	977051390	182.10
MUSTARD, YELLOW, EXTRACT (BRASSICA SPP.)	977091807	182.20
MYRCENE	000123353	172.515
MYRISTALDEHYDE	000124254	172.515
MYRISTIC ACID	000544638	173.340 172.210

MAINTERM	CAS	REGNUM
		172.860
MYRISTYL ALCOHOL	000112721	172.864
MYRRH, GUM (COMMIPHORA SPP.)	009000457	172.510
MYRRH, OIL (COMMIPHORA SPP.)	008016373	172.510
MYRTENOL	000515004	
MYRTENYL ACETATE	001079012	
MYRTLE LEAVES (MYRTUS COMMUNIS L.)	977070851	172.510
NAPHTHA	977126649	73.1
2-NAPHTHALENTHIOL	000091601	
B&-NAPHTHYL ANTHRANILATE	063449683	
B&-NAPHTHYL ETHYL ETHER	000093185	
B&-NAPHTHYL ISOBUTYL ETHER	002173571	
B&-NAPHTHYL METHYL ETHER	000093049	
NARINGIN, EXTRACT (CITRUS PARADISI MACF.)	977038871	182.20
NATAMYCIN	007681938	172.155
NATURAL GAS	008006142	173.350
N-BUTOXYPOLYOXYETHYLENE POLYOXYPROPYLENE GLYCOL	009038953	173.340
D-NEOMENTHOL	002216526	172.515
NEROL	000106252	172.515
NEROLI BIGARADE OIL (CITRUS AURANTIUM L.)	008016384	182.20
NEROLIDOL	007212444	172.515
NERYL ACETATE	000141128	172.515
NERYL BUTYRATE	000999406	172.515
NERYL FORMATE	002142941	172.515
NERYL ISOBUTYRATE	002345246	172.515

MAINTERM	CAS	REGNUM
NERYL ISOVALERATE	003915831	172.515
NERYL PROPIONATE	000105919	172.515
NIACIN	000059676	182.5530 PART 139 PART 137 136.115 184.1530
NIACINAMIDE	000098920	182.5535 184.1535
NICKEL	007440020	172.864 184.1537
NICOTINAMIDE-ASCORBIC ACID COMPLEX	006485445	172.315
NIGER GUTTA (FICUS PLATYPHYLLA DEL.)	977011496	172.615
NISIN PREPARATION	977127335	184.1538
NITRATES, SODIUM & POTASSIUM	977124881	
NITRITES, SODIUM & POTASSIUM	977124892	181.34
NITROGEN	007727379	169.140 169.115 184.1540 169.150
NITROGEN OXIDES	977099254	137.105 137.155 137.165 137.160 137.170 137.175 137.180 137.185
NITROSYL CHLORIDE	002696926	137.200 137.105 137.205 137.155 137.165 137.160 137.170 137.175 137.180 137.185
NITROUS OXIDE	010024972	184.1545
2,4-NONADIENAL	006750034	
2-TRANS-6-CIS-NONADIENAL	000557482	

MAINTERM	CAS	REGNUM
2-TRANS-6-TRANS-NONADIENAL	017587336	
2,6-NONADIENAL DIETHYL ACETAL	106950349	
2,6-NONADIEN-1-OL	007786449	172.515
G&-NONALACTONE	000104610	172.515
NONANAL	000124196	172.515
1,3-NONANEDIOL ACETATE (MIXED ESTERS)	001322174	172.515
1,4-NONANEDIOL DIACETATE	067715815	
1,9-NONANEDITHIOL	003489289	
NONANOIC ACID	000112050	172.515 173.315
2-NONANOL	000628999	
2-NONANONE	000821556	172.515
3-NONANONE	000925780	
3-NONANON-1-OL	067801461	
3-NONANON-1-YL ACETATE	007779546	172.515
NONANOYL 4-HYDROXY-3-METHOXYBENZYLAMIDE	002444464	172.515
2-NONENAL	002463538	
CIS-6-NONENAL	002277192	
CIS-6-NONEN-1-OL	035854865	
TRANS-2-NONEN-1-OL	031502144	
CIS-2-NONEN-1-OL	041453569	
NONYL ACETATE	000143135	172.515
NONYL ALCOHOL	000143088	172.515
NONYL ISOVALERATE	007786472	172.515
NONYL OCTANOATE	007786483	172.515
NOOTKATONE	004674504	172.515

MAINTERM	CAS	REGNUM
NORDIHYDROGUAIARETIC ACID--PROHIBITED	000500389	189.165
NUTMEG (MYRISTICA FRAGRANS HOUTT.)	977051447	182.10
NUTMEG, OIL (MYRISTICA FRAGRANS HOUTT.)	008008455	182.20
NUTMEG OLEORESIN	008007123	182.20
OAK CHIPS, EXTRACT (QUERCUS ALBA L.)	977083138	172.510
OAK MOSS, ABSOLUTE (EVERNIA SPP.)	977059156	172.510
OAK WOOD, ENGLISH (QUERCUS ROBUR L.)	977089909	172.510
OAT GUM	977019241	133.134 133.178 PART 135 133.179
OCIMENE	013877913	172.515
9,12-OCTADECADIENOIC ACID (48%) AND 9,12,15-OCTADECATRIENOIC ACID (52%)	977043767	
OCTADECYLAMINE	000124301	173.310
2-TRANS-6-TRANS-OCTADIENAL	056767181	
TRANS,TRANS-2,4-OCTADIENAL	030361285	
OCTAFLUOROCYCLOBUTANE	000115253	173.360
D&-OCTALACTONE	000698760	
G&-OCTALACTONE	000104507	172.515
OCTANAL	000124130	172.515
OCTANAL DIMETHYL ACETAL	010022283	172.515
1,8-OCTANEDITHIOL	001191624	
OCTANOIC ACID	000124072	172.860 186.1025 184.1025 173.340 172.210
1-OCTANOL	000111875	172.515 172.230

MAINTERM	CAS	REGNUM
2-OCTANOL	000123966	172.515
3-OCTANOL	000589980	172.515
2-OCTANONE	000111137	172.515
3-OCTANONE	000106683	172.515
3-OCTANON-1-OL	007786529	172.515
2-OCTENAL	002363895	
6-OCTENAL	063826255	
CIS-5-OCTENAL	041547222	
CIS-3-OCTEN-1-OL	020125842	
1-OCTEN-3-OL	003391864	172.515
CIS-5-OCTEN-1-OL	064275736	
TRANS-3-OCTEN-2-OL	057648552	
1-OCTEN-3-ONE	004312996	
2-OCTEN-4-ONE	004643270	
3-OCTEN-2-ONE	001669449	
TRANS-2-OCTEN-1-YL ACETATE	003913802	
1-OCTEN-3-YL ACETATE	002442106	172.515
TRANS-2-OCTEN-1-YL BUTANOATE	084642604	
1-OCTEN-3-YL BUTYRATE	016491546	
1-OCTENYL SUCCINIC ANHYDRIDE	007757962	172.892
OCTYL ACETATE	000112141	172.515
3-OCTYL ACETATE	004864613	172.515
OCTYL ALCOHOL, SYNTHETIC	977152923	173.280 172.864 178.3480
OCTYL BUTYRATE	000110394	172.515

MAINTERM	CAS	REGNUM
OCTYL FORMATE	000112323	172.515
OCTYL 2-FUROATE	039251882	
OCTYL GALLATE	001034011	166.110
OCTYL HEPTANOATE	005132752	172.515
OCTYL ISOBUTYRATE	000109159	172.515
OCTYL ISOVALERATE	007786585	172.515
OCTYL 2-METHYLBUTYRATE	029811505	
OCTYL OCTANOATE	002306889	172.515
OCTYL PHENYLACETATE	000122452	172.515
OCTYL PROPIONATE	000142609	172.515
OITICIA OIL	008016351	
OLEIC ACID	000112801	182.90 173.315(A)(3) 172.860 182.70 173.315 172.210
OLEIC ACID, FROM TALL OIL FATTY ACIDS	977047394	173.340 172.862 172.210
OLIBANUM, OIL (BOSWELLIA SPP.)	008016362	172.510
ONION, OIL (ALLIUM CEPA L.)	008002720	182.20
OPOPANAX, GUM	009000786	172.510
OPOPANAX, NON-SPECIFIC	977136063	172.510
OPOPANAX, OIL	008021361	172.510
OPOPANAX TINCTURE	977091818	172.510
ORANGE B	053060701	74.250
ORANGE BLOSSOMS, ABSOLUTE (CITRUS AURANTIUM L.)	977049652	182.20
ORANGE ESSENCE, NATURAL	977091852	
ORANGE ESSENCE OIL, NATURAL	068514750	

MAINTERM	CAS	REGNUM
ORANGE FLOWERS (CITRUS AURANTIUM L.)	977051527	182.20
ORANGE LEAF, ABSOLUTE (CITRUS AURANTIUM L.)	977091841	182.20
ORANGE, OIL, DISTILLED (CITRUS SINENSIS (L.) OSBECK)	977091830	182.20 146.151 146.146
ORANGE, OIL, TERPENELESS (CITRUS SINENSIS (L.) OSBECK)	068606940	
ORANGE PEEL	977070862	182.20
ORANGE PEEL, BITTER, EXTRACT (CITRUS AURANTIUM L.)	977081870	182.20
ORANGE PEEL, BITTER, OIL (CITRUS AURANTIUM L.)	068916041	182.20
ORANGE PEEL, SWEET, EXTRACT (CITRUS SINENSIS (L.) OSBECK)	977091829	182.20
ORANGE PEEL, SWEET, OIL (CITRUS SINENSIS (L.) OSBECK)	008008579	182.20
ORANGE PEEL, SWEET, OIL, TERPENELESS (CITRUS SINENSIS (L.) OSBECK)	977154098	182.20
OREGANO, EUROPEAN (ORIGANUM SPP.)	977002100	
OREGANO (LIPPIA SPP., USUALLY L. GRAVEOLENS HBK)	977138707	182.10
OREGANO (OTHER GENERA INCLUDING COLEUS, LANTANA AND HYPTIS)	977138694	
ORIGANUM OIL (EXTRACTIVE)(THYMUS CAPITATUS HOFF. ET LINK)	008007112	
ORRIS, CONCRETE, LIQUID, OIL (IRIS FLORENTINA L.)	977086433	
ORRIS ROOT, EXTRACT (IRIS FLORENTINA L.)	008002731	172.510
OSMANTHUS ABSOLUTE	977103813	
OX BILE EXTRACT	008008637	184.1560
OXIRANE (CHLOROMETHYL)-, POLYMER WITH AMMONIA, REACTION PRODUCT WITH CHLOROMETHANE	068036997	173.25

MAINTERM	CAS	REGNUM
3-OXOBUTANAL, DIMETHYL ACETAL	005436215	
3-OXODECANOIC ACID GLYCERIDE	977148030	
3-OXODODECANOIC ACID GLYCERIDE	977148041	
3-OXOHEXADECANOIC ACID GLYCERIDE	977148074	
3-OXOHEXANOIC ACID DIGLYCERIDE	977148063	
3-OXOOCTANOIC ACID GLYCERIDE	977148052	
3-OXOTETRADECANOIC ACID GLYCERIDE	977148085	
OXYSTEARIN	008028453	173.340 169.140 169.115 169.150 172.818
P-4000--PROHIBITED	000553797	189.175
PALMITIC ACID	000057103	173.340 172.860 172.210
PANCREATIC EXTRACT	977071810	
PANSY (VIOLA TRICOLOR L.)	977068828	172.510
D-PANTOTHENAMIDE	007757973	172.335
D-PANTOTHENYL ALCOHOL	000081130	182.5580
PAPAIN(CARICA PAPAYA L.)	009001734	137.305 184.1585
PAPRIKA (CAPSICUM ANNUUM L.)	977006453	73.340 182.10
PAPRIKA OLEORESIN (CAPSICUM ANNUUM L.)	068917782	73.345 182.20
PARAFFIN AND SUCCINIC DERIVATIVES, SYNTHETIC	977051572	172.275
PARAFFIN WAX	008002742	172.615 133.181
PARMESAN CHEESE, REGGIANO CHEESE	977090871	133.165
PARSLEY, OIL (PETROSELINUM SPP.)	008000688	182.20

MAINTERM	CAS	REGNUM
PARSLEY, OLEORESIN (PETROSELINUM SPP.)	008025954	182.20
PARSLEY (PETROSELINUM SPP.)	977051583	182.10
PASSION FLOWER EXTRACT	008057623	172.510
PASSION FLOWER (PASSIFLORA INCARNATA L.)	977001538	172.510
PATCHOULY, OIL (POGOSTEMON SPP.)	008014093	172.510
A&-(P-DODECYLPHENYL)-W&-HYDROXYPOL-Y(OXYETHYLENE)	026401478	172.710
PEACH KERNEL, EXTRACT(PRUNUS PERSICA SIEB ET ZUCC.)	008023981	182.40
PEACH LEAVES (PRUNUS PERSICA (L.) BATSCH)	977009838	172.510
PEANUT OIL	008002037	182.70
PEANUT STEARINE (ARACHIS HYPOGAEA L.)	977051594	182.40
PECTIN	009000695	135.140 PART 145 PART 150 173.385 184.1588
PECTINASE FROM ASPERGILLUS NIGER	977031858	
PECTINASE FROM BACILLUS SUBTILIS	977090122	
PECTIN, MODIFIED	977091874	173.385 184.1588
PENDARE (COUMA MACROCARPA BARB. RODR. & COUMA UTILIS (MART.) MUELL. ARG.)	977011509	172.615
PENICILLINASE FROM BACILLUS SUBTILIS	977090133	
PENICILLIUM ROQUEFORTI	977083149	
PENNYROYAL, OIL, AMERICAN (HEDEOMA PULEGIODES (L.))	008007441	172.510
PENNYROYAL, OIL, EUROPEAN (MENTHA PULEGIUM L.)	008013998	172.510
W&-PENTADECALACTONE	000106025	172.515

MAINTERM	CAS	REGNUM
2-PENTADECANONE	002345280	
2,4-PENTADIENAL	000764409	
2,3-PENTANEDIONE	000600146	172.515
2-PENTANOL	006032297	
2-PENTANONE	000107879	172.515
2-PENTENAL	000764396	
4-PENTENOIC ACID	000591800	172.515
1-PENTEN-3-OL	000616251	172.515
2-PENTEN-1-OL	020273249	
1-PENTEN-3-ONE	001629589	
3-PENTEN-2-ONE	000625332	
2-PENTYL-1-BUTEN-3-ONE	063759557	
2-TERT-PENTYLCYCLOHEXYL ACETATE	067874720	
2-PENTYLFURAN	003777693	
PENTYL 2-FURYL KETONE	014360500	
2-PENTYL-3-METHYL-2-CYCLOPENTEN-1--ONE	001128081	
2-PENTYLPYRIDINE	002294760	
PEPPER, BLACK, OIL (PIPER NIGRUM L.)	008006824	182.20
PEPPER, BLACK, OLEORESIN (PIPER NIGRUM L.)	008002560	182.20
PEPPER, BLACK (PIPER NIGRUM L.)	977051629	182.10
PEPPERMINT LEAVES (MENTHA PIPERITA L.)	977018191	182.10 155.130 155.200 155.120 155.170 145.131
PEPPERMINT, OIL (MENTHA PIPERITA L.)	008006904	182.20

MAINTERM	CAS	REGNUM
PEPPERMINT PLANT	977001367	182.10
PEPPER, WHITE, OIL (PIPER NIGRUM L.)	977018204	182.20
PEPPER, WHITE, OLEORESIN (PIPER NIGRUM L.)	977018215	182.20
PEPPER, WHITE (PIPER NIGRUM L.)	977051630	182.10
PEPSIN	009001756	137.305
PEPTONE	977027885	184.1553
PERACETIC ACID	000079210	172.892 172.560 178.1010
PERILLALDEHYDE	002111753	172.515
PERILLYL ACETATE	015111963	172.515
PETITGRAIN, LEMON, OIL (CITRUS LIMON (L.) BURM. F.)	008048519	182.20
PETITGRAIN, MANDARIN, OIL (CITRUS RETICULATA BLANCO VAR. MANDARIN)	977051674	182.20
PETITGRAIN, OIL (CITRUS AURANTIUM L.)	008014173	182.20
PETROLATUM	008009038	173.340 172.880 573.720
PETROLEUM HYDROCARBONS, ISOPARAFFINIC, SYNTHETIC	977051696	173.340 173.280 172.882
PETROLEUM HYDROCARBONS, ODORLESS, LIGHT	977051685	173.340 172.884 573.740
PETROLEUM NAPHTHA	008030306	172.210 172.250
PETROLEUM WAX	977051709	173.340 172.886 172.615 172.230
PETROLEUM WAX, SYNTHETIC	977045730	172.888 172.615 173.340

MAINTERM	CAS	REGNUM
A&-PHELLANDRENE	000099832	172.515
PHENETHYL ACETATE	000103457	172.515
PHENETHYL ALCOHOL	000060128	172.515
PHENETHYLAMINE	000064040	
PHENETHYL ANTHRANILATE	000133186	172.515
PHENETHYL BENZOATE	000094473	172.515
PHENETHYL BUTYRATE	000103526	172.515
PHENETHYL CINNAMATE	000103537	172.515
PHENETHYL FORMATE	000104621	172.515
PHENETHYL 2-FUROATE	007149328	
PHENETHYL HEXANOATE	006290375	
PHENETHYL ISOBUTYRATE	000103480	172.515
PHENETHYL ISOVALERATE	000140261	172.515
PHENETHYL 2-METHYLBUTYRATE	024817514	172.515
PHENETHYL OCTANOATE	005457705	
PHENETHYL PHENYLACETATE	000102205	172.515
PHENETHYL PROPIONATE	000122703	172.515
PHENETHYL SALICYLATE	000087229	172.515
PHENETHYL SENECIOATE	042078659	172.515
PHENETHYL TIGLATE	055719852	172.515
PHENOL	000108952	175.105
		175.300
		175.380
		175.390
		176.170
		177.1210
		177.1580
		177.2410
PHENOL-FORMALDEHYDE, CROSS-LINKED, TETRAETHYLENEPENTAMINE ACTIVATED	027233927	173.25
PHENOL-FORMALDEHYDE, CROSS-LINKED, TRIETHYLENETETRAMINE ACTIVATED	032610778	173.25

MAINTERM	CAS	REGNUM
PHENOL-FORMALDEHYDE, CROSS-LINKED,TRIETHYLENETETRAMINE & TETRAETHYLENEPENTAMINE	977083150	173.25
PHENOL-FORMALDEHYDE, SULFITE-MODIFIED, CROSS-LINKED	977083161	173.25
PHENOXYACETIC ACID	000122598	172.515
2-PHENOXYETHYL ISOBUTYRATE	000103606	172.515
PHENYLACETALDEHYDE	000122781	172.515
PHENYLACETALDEHYDE 2,3-BUTYLENE GLYCOL ACETAL	005468064	172.515
PHENYLACETALDEHYDE DIISOBUTYL ACETAL	068345222	
PHENYLACETALDEHYDE DIMETHYL ACETAL	000101484	172.515
PHENYLACETALDEHYDE GLYCERYL ACETAL	029895736	172.515
PHENYLACETIC ACID	000103822	172.515
DL-PHENYLALANINE	000150301	
L-PHENYLALANINE	000063912	172.320
4-PHENYL-2-BUTANOL	002344709	172.515
2-PHENYL-2-BUTENAL	004411896	
4-PHENYL-3-BUTEN-2-OL	017488652	172.515
4-PHENYL-3-BUTEN-2-ONE	000122576	172.515
4-PHENYL-2-BUTYL ACETATE	010415880	172.515
2-PHENYL-3-CARBETHOXY FURAN	050626023	
PHENYL DISULFIDE	000882337	
2-PHENYL-3-(2-FURYL)-PROP-2-ENAL	057568602	
1-PHENYL-3-METHYL-3-PENTANOL	010415879	172.515
5-PHENYLPENTANOL	010521912	
2-PHENYL-4-PENTENAL	024401363	

MAINTERM	CAS	REGNUM
3-PHENYL-4-PENTENAL	000939219	
1-PHENYL-1,2-PROPANEDIONE	000579077	
1-PHENYL-1-PROPANOL	000093549	172.515
3-PHENYL-1-PROPANOL	000122974	172.515
2-PHENYLPROPIONALDEHYDE	000093538	172.515
3-PHENYLPROPIONALDEHYDE	000104530	172.515
2-PHENYLPROPIONALDEHYDE DIMETHYL ACETAL	000090879	172.515
3-PHENYLPROPIONIC ACID	000501520	172.515
3-PHENYLPROPYL ACETATE	000122725	172.515
2-PHENYLPROPYL BUTYRATE	080866837	172.515
3-PHENYLPROPYL CINNAMATE	000122689	172.515
3-PHENYLPROPYL FORMATE	000104643	172.515
3-PHENYLPROPYL HEXANOATE	006281409	172.515
2-PHENYLPROPYL ISOBUTYRATE	065813538	172.515
3-PHENYLPROPYL ISOBUTYRATE	000103582	172.515
3-PHENYLPROPYL ISOVALERATE	005452073	172.515
3-PHENYLPROPYL PROPIONATE	000122747	172.515
1-PHENYL-3 OR 5-PROPYLPYRAZOLE	065504930	
2-(3-PHENYLPROPYL)PYRIDINE	002110181	
2-(3-PHENYLPROPYL)TETRAHYDROFURAN	003208400	172.515
PHOSPHORIC ACID	007664382	182.1073 131.144 PART 133
PHOSPHORUS OXYCHLORIDE	010025873	172.892
PIMENTA LEAF, OIL (PIMENTA OFFICINALIS LINDL.)	008016453	182.20
PINE BARK, WHITE, EXTRACT SOLID (PINUS STROBUS L.)	977089636	172.510

MAINTERM	CAS	REGNUM
PINE BARK, WHITE, OIL (PINUS STROBUS L.)	977089625	172.510
PINE BARK, WHITE (PINUS STROBUS L.)	977002917	172.510
A&-PINENE	000080568	172.515
B&-PINENE	000127913	172.515
PINE NEEDLE, DWARF, OIL (PINUS MUGO TURRA VAR. PUMILIO (HAENKE) ZENARI)	008000268	172.510
PINE, SCOTCH, OIL (PINUS SYLVESTRIS L.)	008023992	172.510
PINE TAR, OIL (PINUS SPP.)	977009974	172.515
PINE, WHITE, OIL (PINUS SPP.)	977019445	172.510
PINOCARVEOL	005947364	172.515
PIPERAZINE DIHYDROCHLORIDE	000142643	NO LONGER UGRAS
PIPERIDINE	000110894	PART 133 172.515
PIPERINE	000094622	172.515
PIPERITENONE	000491098	172.515
PIPERITENONE OXIDE	035178553	172.515
D-PIPERITONE	006091505	172.515
PIPERONAL	000120570	182.60
PIPERONYL ACETATE	000326614	172.515
PIPERONYL ISOBUTYRATE	005461085	172.515
PIPSISSEWA LEAVES, EXTRACT (CHIMAPHILA UMBELLATA NUTT.)	977023190	182.20
POLYACRYLAMIDE	009003058	173.315 172.255
POLYACRYLAMIDE RESIN, MODIFIED	026006224	173.10
POLY(ACRYLIC ACID-CO-HYPOPHOSPHITE), SODIUM SALT	071050629	173.310
POLYACRYLIC ACID, SODIUM SALT	009003047	173.340 173.310

MAINTERM	CAS	REGNUM
		173.73
POLYDEXTROSE	068424044	172.841
POLYETHYLENE GLYCOL (400) DIOLEATE	977051754	173.340
POLYETHYLENE GLYCOL (M W 200-9,500)	025322683	172.820 173.340 172.210 73.1 173.310
POLYETHYLENE (M W 2,000-21,000)	009002884	172.615 173.20
POLYETHYLENE, OXIDIZED	068441178	172.260
POLYETHYLENIMINE REACTION PRODUCT W/ 1,2-DICHLOROETHANE	068130972	173.357
POLYGLYCEROL ESTERS OF FATTY ACIDS	977050693	169.150 166.110 169.140 172.854 169.115
POLYGLYCERYL PHTHALATE ESTER OF COCONUT OIL FATTY ACIDS	977092946	172.710
POLYISOBUTYLENE (MIN M W 37,000)	009003274	172.615
POLYLIMONENE	009003730	172.515
POLYMALEIC ACID	026099092	173.310 173.45
POLYMALEIC ACID, SODIUM SALT	070247904	173.310 173.45
POLYOXYETHYLENE DIOLEATE	009005076	173.340(A)(3) 173.340
POLYOXYETHYLENE (600) DIOLEATE	977028991	173.340
POLYOXYETHYLENE (600) MONO- RICINOLEATE	977137782	173.340
POLYOXYETHYLENE 40 MONOSTEARATE	009004993	173.340
POLYPROPYLENE GLYCOL (M W 1,200-2,500)	025322694	173.340 173.310
POLYSORBATE 20	009005645	172.515
POLYSORBATE 60	009005678	173.340 172.836

MAINTERM	CAS	REGNUM
		163.130
		163.123
		172.515
		73.1001
		573.840
		163.135
		163.140
		163.145
		163.155
		163.150
		163.153
POLYSORBATE 65	009005714	172.838
		173.340
		73.1001
POLYSORBATE 80	009005656	172.840
		73.1
		172.515
		173.340
		73.1001
		573.860
		172.623
		175.105
		175.300
		176.180
		178.3400
POLYSTYRENE, CROSS-LINKED	977086875	173.25
POLYVINYL ACETATE	009003207	73.1
		172.615
POLYVINYL ALCOHOL	009002895	73.1
POLYVINYL POLYPYRROLIDONE	977043972	173.50
POLYVINYLPYRROLIDONE	009003398	73.1
		172.210
		173.55
		73.1001
POMEGRANATE BARK, EXTRACT (PUNICA GRANATUM L.)	977018226	182.20
POPLAR BUDS (POPULUS SPP.)	977002202	172.510
POPPY SEED (PAPAVER SOMNIFERUM L.)	977051776	182.10
POTASSIUM ACETATE	000127082	172.515
POTASSIUM ACID PYROPHOSPHATE	977050637	
POTASSIUM ACID TARTRATE	000868144	184.1077
		150.141
		150.161
POTASSIUM BENZOATE	000582252	166.110
POTASSIUM BICARBONATE	000298146	184.1613
		163.110

MAINTERM	CAS	REGNUM
POTASSIUM BISULFITE	007773037	182.3616
POTASSIUM BORATE	001332770	
POTASSIUM BROMATE	007758012	172.730 136.110 137.205 137.155 136.115 136.130 136.160 136.180
POTASSIUM BROMIDE	007758023	173.315
POTASSIUM CAPRATE	013040181	172.863
POTASSIUM CAPRYLATE	000764716	172.863
POTASSIUM CARBONATE	000584087	173.310 172.560 184.1619 163.110
POTASSIUM CASEINATE	068131544	135.140 135.110
POTASSIUM CHLORIDE	007447407	166.110 182.5622 150.161 150.141 184.1622
POTASSIUM CITRATE	006100056	182.1625 182.6625 133.179 133.169 150.141 133.173 150.161 184.1625
POTASSIUM CYCLAMATE--PROHIBITED	007758045	189.135
POTASSIUM 2-(1'-ETHOXY)ETHOXYPROPANOATE	100743688	
POTASSIUM FUMARATE	007704725	172.350
POTASSIUM GIBBERELLATE	000125677	172.725
POTASSIUM GLUCONATE	000299274	
POTASSIUM GLYCEROPHOSPHATE	001319706	182.5628 182.8628
POTASSIUM HYDROXIDE	001310583	184.1631 163.110

MAINTERM	CAS	REGNUM
POTASSIUM HYPOPHOSPHATE	977052519	
POTASSIUM HYPOPHOSPHITE	007782878	
POTASSIUM IODATE	007758056	582.80 136.110 184.1635 136.115 136.130 136.160 136.180
POTASSIUM IODIDE	007681110	184.1634 582.80 172.375
POTASSIUM LACTATE	000996316	184.1639
POTASSIUM LAURATE	010124659	172.863
POTASSIUM METABISULFITE	016731558	182.3637
POTASSIUM N-METHYLDITHIOCARBAMATE	000137417	173.320
POTASSIUM MYRISTATE	013429271	172.863
POTASSIUM NITRATE	007757791	172.160 181.33
POTASSIUM NITRITE	007758090	181.34
POTASSIUM OLEATE	000143180	172.863
POTASSIUM PALMITATE	002624319	172.863
POTASSIUM PERMANGANATE	007722647	172.892
POTASSIUM PERSULFATE	007727211	172.210
POTASSIUM PHOSPHATE, DIBASIC	007758114	133.169 182.6285 133.173 133.179
POTASSIUM PHOSPHATE, MONOBASIC	007778770	160.110
POTASSIUM PHOSPHATE, TRIBASIC	007778532	175.105 176.170 176.180
POTASSIUM POLYMETAPHOSPHATE	007790536	
POTASSIUM PYROPHOSPHATE	007320345	173.315(A)(3) 173.315

MAINTERM	CAS	REGNUM
POTASSIUM SALTS OF FATTY ACIDS	977093392	172.863
POTASSIUM SORBATE	000590001	182.3640 182.90 166.110 150.141 PART 133 150.161
POTASSIUM STEARATE	000593293	172.615 172.863 173.340
POTASSIUM SULFATE	007778805	184.1643
POTASSIUM TRIPOLYPHOSPHATE	013845368	173.310
PRICKLY ASH BARK EXTRACT (XANTHOXYLUM SPP.)	977009996	182.20
PRICKLY ASH BARK, OIL	977018248	182.20
L-PROLINE	000147853	172.320
PROPANE	000074986	184.1655 173.350
1,2-PROPANEDITHIOL	000814675	
1,3-PROPANEDITHIOL	000109808	
4-PROPENYL-2,6-DIMETHOXYPHENOL	006635229	
PROPENYLGUAETHOL	000094860	172.515
PROPENYL PROPYL DISULFIDE	005905464	
PROPIONALDEHYDE	000123386	172.515
PROPIONIC ACID	000079094	184.1081
2-PROPIONYLPYRROLE	001073263	
2-PROPIONYLTHIAZOLE	043039981	
PROPIOPHENONE	000093550	
PROPYL ACETATE	000109604	172.515
PROPYL ALCOHOL	000071238	172.515 573.880
P-PROPYLANISOLE	000104450	172.515

MAINTERM	CAS	REGNUM
PROPYL BENZOATE	002315686	172.515
PROPYL BUTYRATE	000105668	172.515
PROPYL CINNAMATE	007778838	172.515
PROPYL 2,4-DECADIENOATE	084788089	
4-PROPYL-2,6-DIMETHOXYPHENOL	006766821	
PROPYL DISULFIDE	000629196	172.515
PROPYLENE CHLOROHYDRIN	000078897	172.892
PROPYLENE GLYCOL	000057556	184.1666 582.4666 169.175 169.176 169.177 169.178 169.180 169.181
PROPYLENE GLYCOL ALGINATE	009005372	173.340 172.858 133.162 172.210 133.178 133.179 133.134 133.133
PROPYLENE GLYCOL DIBENZOATE	019224261	
PROPYLENE GLYCOL MONO- AND DIESTERS OF FATTY ACIDS	977050706	173.340 136.110 172.856 136.115 136.130 136.160 136.180
PROPYLENE GLYCOL STEARATE	001323393	
PROPYLENE OXIDE	000075569	172.892 175.105 176.210 178.3120 193.380
PROPYL FORMATE	000110747	172.515
PROPYL 2-FURANACRYLATE	000623223	172.515
PROPYL 2-FUROATE	000615101	
PROPYL GALLATE	000121799	166.110 172.615

MAINTERM	CAS	REGNUM
		184.1660
		175.125
		175.300
		175.380
		175.390
		176.170
		177.1010
		177.1210
		177.1350
PROPYL HEPTANOATE	007778872	172.515
PROPYL HEXANOATE	000626777	172.515
PROPYL P-HYDROXYBENZOATE	000094133	150.161
		184.1670
		172.515
		150.141
3-PROPYLIDENEPHTHALIDE	017369594	172.515
PROPYL ISOBUTYRATE	000644495	172.515
PROPYL ISOVALERATE	000557006	172.515
PROPYL MERCAPTAN	000107039	172.515
PROPYL 2-METHYL-3-FURYL DISULFIDE	061197099	
A&-PROPYLPHENETHYL ALCOHOL	000705737	172.515
O-PROPYLPHENOL	000644359	
P-PROPYLPHENOL	000645567	
PROPYL PHENYLACETATE	004606159	172.515
PROPYL PROPIONATE	000106365	172.515
PROPYL THIOACETATE	002307100	
PROTEASE FROM ASPERGILLUS FLAVUS	977017314	
PROTEASE FROM ASPERGILLUS NIGER	977031927	
PROTEASE FROM ASPERGILLUS ORYZAE	977017336	
PROTEASE FROM BACILLUS LICHENIFORMIS	977083172	
PROTEASE FROM BACILLUS SUBTILIS	051931238	
PROTEIN, ANIMAL, HYDROLYZED	100085618	
PROTEIN HYDROLYSATE	009015547	161.190(A)(6)

MAINTERM	CAS	REGNUM
		573.200
PROTEIN, MILK, HYDROLYZED	092797392	
PROTEIN, VEGETABLE, HYDROLYZED	100209458	155.170
		155.130
		155.200
		155.120
		145.131
PSYLLIUM SEED HUSK	977022222	PART 135
A&-(P-(1,1,3,3-TETRAMETHYLBUTYL)PH-ENYL)-W&-HYDROXYPOLY(OXYETHYLENE)(1 MOL)	002315675	172.710
PULEGONE	000089827	172.515
PULPS FROM WOOD, STRAW, BAGASSE, OR OTHER NATURAL SOURCES	888285845	186.1673
PYRAZINE ETHANETHIOL	035250534	
PYRAZINYL METHYL SULFIDE	021948709	
PYRIDINE	000110861	172.515
		177.1580
2,6-PYRIDINEDICARBOXYLIC ACID	000499832	178.1010
2-PYRIDINEMETHANETHIOL	002044737	
PYRIDOXINE	000065236	
PYRIDOXINE HYDROCHLORIDE	000058560	182.5676
		184.1676
PYROLIGNEOUS ACID	008030975	
PYROLIGNEOUS ACID, EXTRACT	008028475	172.515
PYRROLE	000109977	
PYRROLIDINE	000123751	
PYRUVALDEHYDE	000078988	172.515
PYRUVIC ACID	000127173	172.515
QUASSIA, EXTRACT (PICRASMA EXCELSA (SW.) PLANCH OR QUASSIA AMARA L.)	068915322	172.510
QUATERNARY AMMONIUM CHLORIDE COMBINATION	977127824	172.165
		173.320

MAINTERM	CAS	REGNUM
QUEBRACHO BARK EXTRACT	977092719	172.510
QUILLAIA (QUILLAJA SAPONARIA MOLINA)	977002279	172.510
QUINCE SEED, EXTRACT (CYDONIA SPP.)	977018259	182.40
QUININE BISULFATE	000549564	
QUININE HYDROCHLORIDE	000130892	172.575
QUININE SULFATE	006119706	172.575
QUINOLINE	000091225	
RAPESEED OIL, HYDROGENATED	084681710	184.1555
RAPESEED OIL, HYDROGENATED, SUPERGLYCERINATED	977011929	184.1555
RAPESEED OIL, LOW ERUCIC ACID	120962030	184.1555
RAPESEED OIL, LOW ERUCIC ACID, PARTIALLY HYDROGENATED	977106356	184.1555
RENNET	009042084	131.187 184.1685 PART 133 131.162 131.160 131.185
RESIN, FROM FORMALDEHYDE, ACETONE, AND TETRAETHYLENEPENTAMINE	009006706	173.25
RESORCINOL	000108463	177.1210
L-RHAMNOSE	003615416	
RHATANY, EXTRACT (KRAMERIA SPF.)	977023634	172.510
RHODINOL	000141253	172.515
RHODINYL ACETATE	000141117	172.515
RHODINYL BUTYRATE	000141151	172.515
RHODINYL FORMATE	000141093	172.515
RHODINYL ISOBUTYRATE	000138238	172.515
RHODINYL ISOVALERATE	007778963	172.515

MAINTERM	CAS	REGNUM
RHODINYL PHENYLACETATE	010486143	172.515
RHODINYL PROPIONATE	000105895	172.515
RHUBARB, GARDEN ROOT (RHEUM RHAPONTICUM L.)	977035941	172.510
RHUBARB ROOT (RHEUM SPP.)	977039943	172.510
RIBOFLAVIN	000083885	182.5695 73.450 PART 137 136.115 PART 139 184.1695
RIBOFLAVIN 5-PHOSPHATE	000146178	182.5697 184.1697
RICE BRAN WAX	008016602	172.890 172.615
RICE, MILLED	977083183	133.153
ROSE, ABSOLUTE (ROSA SPP.)	977091932	182.20
ROSE, BUD (ROSA SPP.)	977029676	182.20
ROSE, BULGARIAN, TRUE OTTO, OIL (ROSA DAMASCENA MILL.)	008007010	182.20
ROSE FLOWERS (ROSA SPP.)	977029698	182.20
ROSE HIPS, EXTRACT (ROSA SPP.)	977021376	182.20
ROSE LEAVES (ROSA SPP.)	977029701	182.20
ROSELLE (HIBISCUS SABDARIFFA L.)	977017881	172.510
ROSEMARY, OIL (ROSEMARINUS OFFICINALIS L.)	008000257	182.20
ROSEMARY, OLEORESIN	977029687	182.20
ROSEMARY (ROSEMARINUS OFFICINALIS L.)	977002360	182.10
ROSE WATER, STRONGER (ROSA CENTIFOLIA L.)	008030260	
ROSIDINHA (MICROPHOLIS (ALSO KNOWN AS SIDEROXYLON) SPP.)	977011510	172.615
ROSIN, ADDUCT WITH FUMARIC ACID, PENTAERYTHRITOL ESTER	065997117	73.1

MAINTERM	CAS	REGNUM
ROSIN, GUM, GLYCEROL ESTER	977035485	172.615
ROSIN, GUM OR WOOD, PENTAERYTHRITOL ESTER	977045810	172.615
ROSIN, GUM OR WOOD, PARTIALLY HYDROGENATED, GLYCEROL ESTER	977074364	172.615
ROSIN, GUM OR WOOD, PARTIALLY HYDROGENATED, PENTAERYTHRITOL ESTER	977045821	172.615
ROSIN, LIMED	009007130	73.1
ROSIN, METHYL ESTER, PARTIALLY HYDROGENATED	977035883	172.615 172.515
ROSIN, PARTIALLY DIMERIZED, CALCIUM SALT	977051867	172.210
ROSIN, PARTIALLY DIMERIZED, GLYCEROL ESTER	977013721	172.615
ROSIN, PARTIALLY HYDROGENATED	977051878	172.210
ROSIN (PINUS SPP.) AND ROSIN DERIVATIVES	008050097	172.615 172.210 73.1 172.510
ROSIN, POLYMERIZED, GLYCEROL ESTER	068475376	172.615
ROSIN, TALL OIL, GLYCEROL ESTER	977019978	172.615
ROSIN, WOOD	009014635	172.210
ROSIN, WOOD, GLYCEROL ESTER	008050304	172.615 172.735
ROSIN, WOOD, MALEIC ANHYD. MOD., PENTAERYTHRITOL ESTER, ACID #134-145	977045978	172.210
ROSIN, WOOD, MALEIC ANHYD. MOD., PENTAERYTHRITOL ESTER, ACID #176-186	977045989	172.210
RUBBER, NATURAL-SMOKED SHEET AND LATEX SOLIDS (HEVEA BRASILIENSIS)	009006046	172.615
RUE, OIL (RUTA GRAVEOLENS L.)	008014297	184.1699
RUE (RUTA GRAVEOLENS L.)	977051889	184.1698

MAINTERM	CAS	REGNUM
RUM	977089454	
RUM ETHER	008030895	172.515
RUTIN	000153184	
SACCHARIN	000081072	150.141 180.37 150.161 145.116 145.126 145.131 145.136 145.171 145.181
SACCHARIN, AMMONIUM SALT	006381619	180.37
SACCHARIN, CALCIUM SALT	006381915	150.141 150.161 180.37
SACCHARIN, SODIUM SALT	000128449	145.116 150.141 150.161 180.37 145.126 145.131 145.136 145.171 145.181
SAFFRON (CROCUS SATIVUS L.)	000138556	182.10 73.500
SAFFRON, EXTRACT (CROCUS SATIVUS L.)	084604171	182.20
SAFROLE--PROHIBITED	000094597	189.180
SAFROLE-FREE EXTRACT OF SASSAFRAS	977051970	172.580
SAGE, GREEK (SALVIA TRILOBA L.)	977051958	182.10
SAGE, OIL (SALVIA OFFICINALIS L.)	008022568	182.20
SAGE, OLEORESIN (SALVIA OFFICINALIS L.)	977029665	182.20
SAGE (SALVIA OFFICINALIS L.)	977002440	182.10
SAGE, SPANISH, OIL (SALVIA LAVANDULAEFOLIA VAHL.)	977125771	182.20
SALICYLALDEHYDE	000090028	172.515
SALTS OF FATTY ACIDS	977009485	172.863

MAINTERM	CAS	REGNUM
SANDALWOOD, RED (PTEROCARPUS SANTALINUS L.F.)	977029712	172.510
SANDALWOOD, WHITE (SANTALUM ALBUM L.)	977020851	172.510
SANDALWOOD, YELLOW, OIL (SANTALUM ALBUM L.)	008006879	172.510
SANDARAC (TETRACLINIS ARTICULATA (VAHL.) MAST.)	009000571	172.510
SANTALOL, A&	000115719	172.515
SANTALOL (A& AND B&)	011031451	172.515
SANTALOL, B&	000077429	172.515
SANTALYL ACETATE	001323008	172.515
SANTALYL PHENYLACETATE	001323757	172.515
SARSAPARILLA, EXTRACT (SMILAX SPP.)	977022675	172.510
SASSAFRAS BARK, EXTRACT (SAFROLE-FREE) (SASSAFRAS ALBIDUM (NUTT.) NEES)	977075232	172.580
SASSAFRAS LEAVES (SAFROLE-FREE) (SASSAFRAS ALBIDUM (NUTT.) NEES)	977088382	172.510
SAUSAGE CASINGS (HCL AND CELLULOSE FIBERS)	888285969	
SAVORY, SUMMER, OIL (SATUREJA HORTENSIS L.)	008016680	182.20
SAVORY, SUMMER, OLEORESIN (SATUREJA HORTENSIS L.)	977029756	182.20
SAVORY, SUMMER (SATUREJA HORTENSIS L.)	977051981	182.10
SAVORY, WINTER, OIL (SATUREJA MONTANA L.)	977029745	182.20
SAVORY, WINTER, OLEORESIN (SATUREJA MONTANA L.)	977029767	182.20
SAVORY, WINTER (SATUREJA MONTANA L.)	977051992	182.10
SCHINUS MOLLE, OIL (SCHINUS MOLLE L.)	068917522	182.20
SENNA, ALEXANDRIA (CASSIA		

MAINTERM	CAS	REGNUM
ACUTIFOLIA DELILE)	977083194	172.510
L-SERINE	000056451	172.320
SERPENTARIA (ARISTOLOCHIA SERPENTARIA L.)	977002553	172.510
SESAME (SESAMUM INDICUM L.)	977052019	182.10
SHELLAC, PURIFIED	009000593	73.1 175.380 175.390 27 CFR 212.61 27 CFR 219.90 40 CFR 180.1001 182.99 175.300(B)(3) 175.105 175.300
SHELLAC WAX	097766502	
SILICON DIOXIDE	007631869	182.90 172.480 173.340 172.230 160.105 133.146 160.185 73.1 573.940
SILVER FIR, NEEDLES AND TWIGS, OIL (ABIES ALBA MILL.)	008021270	172.510
SILVER-SILVER DRAGEES	888286008	
SIMARUBA BARK (SIMARUBA AMARA AUBL.)	977029609	172.510
SKATOLE	000083341	172.515
SLOE BERRIES, EXTRACT (PRUNUS SPINOSA L.)	977029621	182.20
SLOE BERRIES, EXTRACT SOLID (PRUNUS SPINOSA L.)	977029610	182.20
SLOE BERRIES (PRUNUS SPINOSA L.)	977052031	182.20
SNAKEROOT, CANADIAN, OIL (ASARUM CANADENSE L.)	008016691	172.510
SODIUM ACETATE	977127846	182.70 150.141 150.161 173.310 184.1721
SODIUM ACID CITRATE	000144332	

MAINTERM	CAS	REGNUM
SODIUM ACID PYROPHOSPHATE	007758169	133.179 133.169 133.173 137.180 182.1087 161.190
SODIUM N-ALKYLBENZENESULFONATE	977052597	173.315
SODIUM ALUMINATE	001302427	173.310 182.90
SODIUM ALUMINUM PHOSPHATE, ACIDIC OR BASIC	007785888	182.90 133.173 137.180 137.105 133.179 137.270 133.169 182.1781 137.155 137.165 137.160 137.170 137.175 137.180 137.185
SODIUM ALUMINUM SILICATE	001344009	160.105 133.146 160.185 182.2727
SODIUM ASCORBATE	000134032	182.3731
SODIUM BENZOATE	000532321	150.161 184.1733 150.141 166.110 166.40 146.154 146.152 181.23
SODIUM BICARBONATE	000144558	184.1736 137.180 163.110 173.385 137.270
SODIUM BISULFITE	007631905	173.310 182.3739 161.173
SODIUM BORATE	001330434	
SODIUM BOROHYDRIDE	016940662	172.560
SODIUM CALCIUM ALUMINOSILICATE, HYDRATED	001344021	182.2729

MAINTERM	CAS	REGNUM
SODIUM CAPRATE	001002626	172.863
SODIUM CAPRYLATE	001984061	172.863
SODIUM CARBONATE	005968116	173.310
		184.1742
		163.110
SODIUM CASEINATE	009005463	182.1748
		166.110
		135.110
		135.140
SODIUM CHLORIDE	007647145	182.1
		133.124
		131.187
		133.123
		182.90
		182.70
		133.169
		166.110
		133.173
		131.185
		131.111
		131.112
		131.136
		131.144
		131.146
		131.162
		131.170
		133.179
		133.187
		133.188
		133.189
		133.190
		133.195
		136.110
		145.110
		145.130
		PART 155
		PART 156
		PART 158
		161.170
		161.173
		161.190
		163.111
		163.112
		163.113
		163.123
		163.130
		169.115
		169.140
		169.150
		131.138
		136.115
		136.130
		136.160
		136.180
		163.113
		163.114
		163.117
		163.135
		163.140
		163.145
		163.155
		163.150
		163.153
SODIUM CHLORITE	007758192	172.892
		186.1750

MAINTERM	CAS	REGNUM
SODIUM CITRATE	000068042	182.1751 182.6751 133.169 150.141 150.161 131.185 131.160 133.179 131.111 131.112 131.144 133.173 131.138 131.146 184.1751
SODIUM CYCLAMATE--PROHIBITED	000139059	189.135
SODIUM DECYLBENZENESULFONATE	001322981	172.210
SODIUM DEHYDROACETATE	004418262	172.130
SODIUM DIACETATE	000126965	184.1754
SODIUM DIMETHYLDITHIOCARBAMATE	000128041	173.320
SODIUM DITHIONITE	007775146	182.90
SODIUM DODECYLBENZENESULFONATE	025155300	173.315
SODIUM ERYTHORBATE	006381777	
SODIUM 2-ETHYLHEXYL SULFATE	000126921	173.315
SODIUM FERRICITROPYROPHOSPHATE	001332963	
SODIUM FERRITRIPOLYPHOSPHATE	977127948	
SODIUM FLUORIDE	007681494	175.105
SODIUM FORMATE	000141537	186.1756
SODIUM FUMARATE	007704736	172.350
SODIUM GLUCOHEPTONATE	031138655	173.310
SODIUM GLUCONATE	000527071	182.6757
SODIUM HUMATE	068131044	173.310
SODIUM HYDROXIDE	001310732	173.310 184.1763 163.110 172.560 172.892 172.814

MAINTERM	CAS	REGNUM
SODIUM HYPOCHLORITE	007681529	173.315 172.892
SODIUM HYPOPHOSPHITE	007681530	184.1764
SODIUM LACTATE	000072173	184.1768
SODIUM LAURATE	000629254	172.863
SODIUM LAURYL SULFATE	000151213	172.210 172.822
SODIUM LIGNOSULFONATE	008061516	173.310
SODIUM METABISULFITE	007681574	182.3766 173.310
SODIUM METAPHOSPHATE	050813166	182.6760 182.90 133.169 172.892 173.310 182.6769 133.179 169.115 133.173 150.161 150.141
SODIUM METASILICATE	006834920	173.310 184.1769A
SODIUM 2-(4-METHOXYPHENOXY)PROPANOATE	977148096	
SODIUM METHYL SULFATE	000512425	173.385
SODIUM MONO- AND DIMETHYL NAPHTHALENE SULFONATES	977052100	173.315 172.824
SODIUM MYRISTATE	000822128	172.863
SODIUM NITRATE	007631994	173.310 172.170 181.33
SODIUM NITRITE	007632000	172.175 172.177 573.700 181.34
SODIUM OLEATE	000143191	186.1770 172.863
SODIUM PALMITATE	000408355	172.863 186.1771

MAINTERM	CAS	REGNUM
SODIUM PANTOTHENATE	000867812	182.5772
SODIUM PECTINATE	009005598	
SODIUM PHOSPHATE, DIBASIC	007558794	182.1778
		135.110
		182.6290
		182.6778
		173.310
		182.5778
		139.110
		150.141
		133.169
		150.161
		133.179
		133.173
		137.305
		139.115
		139.117
		139.135
		182.8778
SODIUM PHOSPHATE, MONOBASIC	007558807	182.1778
		182.6085
		173.310
		182.6778
		182.5778
		150.141
		160.110
		133.169
		133.179
		163.123
		163.130
		133.173
		172.892
		150.161
		182.8778
		163.135
		163.140
		163.145
		163.155
		163.150
		163.153
SODIUM PHOSPHATE, TRIBASIC	007601549	182.6778
		182.1778
		150.161
		133.179
		133.173
		133.169
		150.141
		173.310
		182.5778
		182.8778
SODIUM POLYMETHACRYLATE	054193361	173.310
SODIUM POTASSIUM TARTRATE	000304596	184.1804
		133.169
		133.173
		133.179
		150.161
		150.141
SODIUM PROPIONATE	000137406	184.1784
		150.161
		133.179
		133.123
		133.124

MAINTERM	CAS	REGNUM
		150.141
		133.169
		133.173
SODIUM PYROPHOSPHATE	007722885	133.173
		133.169
		133.179
		182.6789
		173.310
		182.6787
		182.70
SODIUM SALTS OF FATTY ACIDS	977038155	172.863
SODIUM SESQUICARBONATE	000533960	184.1792
SODIUM SILICATE	001344098	182.70
		182.90
		173.310
SODIUM SORBATE	007757815	182.3795
		150.161
		PART 133
		150.141
		182.90
		166.110
SODIUM STEARATE	000822162	172.615
		172.863
		181.29
SODIUM STEAROYL-2-LACTYLATE	025383997	172.846
SODIUM STEARYL FUMARATE	004070808	172.826
SODIUM SULFATE	007727733	172.615
		173.310
		186.1797
SODIUM SULFIDE	001313822	172.615
SODIUM SULFITE	007757837	173.310
		182.3798
SODIUM TARTRATE	000868188	184.1801
		133.173
		133.179
		150.141
		133.169
		150.161
SODIUM TAUROCHOLATE	000145426	
SODIUM THIOSULFATE	010102177	184.1807
SODIUM TRIPOLYPHOSPHATE	007758294	172.892
		182.90
		182.70
		173.310
		182.6810
		182.1810

MAINTERM	CAS	REGNUM
SODIUM ZINC METASILICATE	090268034	
SORBIC ACID	000110441	146.154 PART 133 150.161 182.3089 146.115 150.141 166.110 146.152
SORBITAN MONOOLEATE	001338438	178.3400 73.1001 173.75
SORBITAN MONOSTEARATE	001338416	173.340 172.842 163.123 163.130 172.515 73.1001 573.960 163.135 163.140 163.145 163.155 163.150 163.153
D-SORBITOL	000050704	182.90 184.1835
SORBOSE	000087796	186.1839
SOYA BEAN OIL FATTY ACIDS, HYDROXYLATED	977038882	173.340
SOYA FATTY ACID AMINE, ETHOXYLATED	061791240	
SOYBEAN OIL, HYDROGENATED	008016704	182.70
SOY CONCENTRATE, ENZYME ACTIVATED	888286031	
SOY PROTEIN, ISOLATES	977076848	166.110 182.90
SPEARMINT, EXTRACT (MENTHA SPICATA L.)	977060175	182.20
SPEARMINT (MENTHA SPICATA L.)	977002611	182.10
SPEARMINT, OIL (MENTHA SPICATA L.)	008008795	182.20
SPERM OIL	008002242	172.210
SPERM OIL, HYDROGENATED	008016737	173.275
SPIKENARD EXTRACT	977071525	

MAINTERM	CAS	REGNUM
SPIRO(2,4-DITHIA-1-METHYL-8-OXABIC- YCLO(3.3.0)OCTANE-3,3'-(1'-OXA-2'-- METHYL)CYCLOPENTANE)	038325256	
SPRUCE NEEDLES AND TWIGS, EXTRACT (PICEA SPP.)	977062739	172.510
SPRUCE NEEDLES AND TWIGS, OIL (PICEA SPP.)	008008808	172.510
STANNIC CHLORIDE	007646788	
STANNOUS CHLORIDE	007772998	155.200 172.180 184.1845
STARCH, ACID MODIFIED	977050740	182.90 172.892
STARCH, BLEACHED	977075425	172.892
STARCH, FOOD, MODIFIED	977052188	172.892 169.179 169.150 169.182
STARCH, FOOD, MODIFIED: ACETYLATED DISTARCH ADIPATE	063798356	172.892 172.892(D)
STARCH, FOOD, MODIFIED: ACETYLATED DISTARCH GLYCEROL	053123845	172.892(F) 172.892
STARCH, FOOD, MODIFIED: ACETYLATED DISTARCH PHOSPHATE	068130143	172.892 172.892(D)
STARCH, FOOD, MODIFIED: DISTARCH GLYCEROL	058944891	172.892(E) 172.892
STARCH, FOOD, MODIFIED: DISTARCH OXYPROPANOL	977043574	172.892(E) 172.892
STARCH, FOOD, MODIFIED: DISTARCH PHOSPHATE (FROM PHOSPHORUS OXYCHLORIDE)	977088757	172.892 172.892(D)
STARCH, FOOD, MODIFIED: DISTARCH PHOSPHATE (FROM SODIUM TRIMETAPHOSPHATE)	977088746	172.892 172.892(D)
STARCH, FOOD, MODIFIED: HYDROXYPROPYL DISTARCH GLYCEROL	059419602	172.892(E) 172.892

MAINTERM	CAS	REGNUM
STARCH, FOOD, MODIFIED: HYDROXYPROPYL DISTARCH PHOSPHATE	053124008	172.892(F) 172.892
STARCH, FOOD, MODIFIED: HYDROXYPROPYL STARCH	009049767	172.892(E) 172.892
STARCH, FOOD, MODIFIED: OXIDIZED HYDROXYPROPYL STARCH	068412862	172.892(G) 172.892
STARCH, FOOD, MODIFIED: OXIDIZED STARCHES	065996625	172.892
STARCH, FOOD, MODIFIED: PHOSPHATED DISTARCH PHOSPHATE	977043585	172.892 172.892(D)
STARCH, FOOD, MODIFIED: STARCH ACETATE	009045287	172.892 172.892(D)
STARCH, FOOD, MODIFIED: STARCH ALUMINUM OCTENYL SUCCINATE	009087610	172.892(D) 172.892
STARCH, FOOD, MODIFIED: STARCH PHOSPHATE	011120028	172.892(D) 172.892
STARCH, FOOD, MODIFIED: STARCH SODIUM OCTENYL SUCCINATE	066829296	172.892(D) 172.892
STARCH, FOOD, MODIFIED: STARCH SODIUM SUCCINATE	037231928	172.892(D) 172.892
STARCH, FOOD, MODIFIED: SUCCINYL DISTARCH GLYCEROL	977043596	172.892(F) 172.892
STARCH, PREGELATINIZED	977050933	172.892(G) 182.90 172.892
STARCH, UNMODIFIED	977052177	182.90 137.105 182.70 169.150 169.179 137.155 137.165 137.160 137.170 137.175 137.180 137.185 169.182

MAINTERM	CAS	REGNUM
STEARIC ACID	000057114	173.340 184.1090 172.860 172.210 172.615
STEARYL ALCOHOL	000112925	172.755 172.864
STEARYL ALCOHOL, PLUS BEESWAX	888286064	
STEARYL CITRATE	001337333	182.6851 166.110 166.40 184.1851
STEARYL MONOGLYCERIDYL CITRATE	001337344	172.755
ST. JOHNSWORT LEAVES, FLOWERS AND CAULIS (HYPERICUM PERFORATUML.)	977092968	172.510
STORAX EXTRACT (LIQUIDAMBAR SPP.)	977029803	172.510
STORAX (LIQUIDAMBAR SPP.)	008046193	172.510
STORAX OIL	008024019	172.510
STYRENE	000100425	173.20 172.515 173.25 175.105 175.300 175.380 175.390 176.170 176.180 177.1010 177.1210 177.2260 177.2420 178.3790
STYRENE-DIVINYLBENZENE-ACRYLONITRI- LE, SULFONATED TERPOLYMER	977089341	173.25(A)(11) 173.25
STYRENE-DIVINYLBENZENE COPOLYMER, CHLOROMETHYLATED, AMINATED, OXIDIZED	977089330	173.25(A)(17) 173.25
STYRENE-DIVINYLBENZENE-METHYL ACRYLATE, SULFONATED TERPOLYMER	977089352	173.25(A)(11) 173.25
STYRENE, DIVINYLBENZENE, SULFONATED COPOLYMER	068037263	173.25(A)(1) 173.25
STYRENE-DVB-ACRYLONITRILE-METHYL ACRYLATE, SULFONATED TETRAPOLYMER	977086886	173.25(A)(15) 173.25

MAINTERM	CAS	REGNUM
SUCCINIC ACID	000110156	184.1091
		131.144
SUCCINIC ANHYDRIDE	000108305	172.892
		175.300
		175.380
		175.390
		176.170
		177.1210
SUCCINYLATED GELATIN	068915242	172.230
SUCCINYLATED MONOGLYCERIDES	977009452	172.830
SUCCISTEARIN	027216622	172.765
SUCROSE	000057501	184.1854
SUCROSE ACETATE ISOBUTYRATE	000126136	175.105
SUCROSE FATTY ACID ESTERS	977019376	172.859
SUCROSE LIQUID	977143795	
SUCROSE OCTAACETATE	000126147	172.515
SUGAR BEET EXTRACT FLAVOR BASE	008016793	172.585
SUGAR SOLID EXTRACT	977029836	
SULFAMIC ACID	005329146	186.1093
SULFITES, STRONG ALKALI	888286133	
SULFITING AGENTS	977116601	182.3616
		182.3637
		182.3739
		182.3766
		182.3798
		182.3862
SULFUR DIOXIDE	007446095	182.3862
		172.892
SULFURIC ACID	007664939	184.1095
		172.892
		172.560
		173.385
		178.1010
TAGETES MEAL & EXTRACT	977010562	73.295
TAGETES, OIL (TAGETES SPP.)	008016840	172.510
TALC	014807966	182.90

MAINTERM	CAS	REGNUM
		182.70
		175.300
		175.380
		175.390
		176.170
		177.1210
		177.1350
		177.1460
TALL OIL	008002264	186.1557
TALLOW ALCOHOL, HYDROGENATED	008030113	173.340
TALLOW, BEEF	061789977	182.70
TALLOW, HYDROGENATED	008030124	182.70
TALLOW, HYDROGENATED, OXIDIZED OR SULFATED	977043643	173.340
TAMARIND EXTRACT (TAMARINDUS INDICA L.)	977029778	182.20
TAMARINDS	977052304	182.20
TANGERINE, ESSENCE	977029789	182.20
TANGERINE, OIL (CITRUS RETICULATA BLANCO)	008016851	182.20
TANNIC ACID	001401554	184.1097
		173.310
TANSY, OIL (TANACETUM VULGARA L.)	008016873	172.510
TANSY (TANACETUM VULGARA L.)	977032442	172.510
TARRAGON (ARTEMISIA DRACUNCULUS L.)	977052326	182.10
TARRAGON EXTRACT (ARTEMISIA DRACUNCULUS L.)	977029814	182.20
TARRAGON OIL (ARTEMISIA DRACUNCULUS L.)	008016884	182.20
TARTARIC ACID	000526830	184.1099
		150.161
		150.141
		131.144
TAURINE	000107357	
TAUROCHOLIC ACID	000081243	
TEA EXTRACT (THEA SINENSIS L.)	084650602	182.20

MAINTERM	CAS	REGNUM
TEA TREE OIL (MELALEUCA ALTERNIFOLIA)	068647734	
TERPENE RESIN	009003741	172.280
TERPENE RESINS, NATURAL	977092253	73.1 172.615 172.280
TERPENE RESINS, SYNTHETIC	977092264	172.615 73.1
A&-TERPINENE	000099865	172.515
G&-TERPINENE	000099854	172.515
A&-TERPINEOL	000098555	172.515
B&-TERPINEOL	000138874	172.515
TERPINOLENE	000586629	172.515
TERPINYL ACETATE	000080262	172.515
A&-TERPINYL ANTHRANILATE	014481528	172.515
TERPINYL BUTYRATE	002153288	172.515
TERPINYL CINNAMATE	010024563	172.515
TERPINYL FORMATE	002153266	172.515
TERPINYL ISOBUTYRATE	007774654	172.515
TERPINYL ISOVALERATE	001142854	172.515
TERPINYL PROPIONATE	000080273	172.515
D&-TETRADECALACTONE	002721224	
TETRAETHYLENEPENTAMINE CROSSLINKED WITH EPICHLOROHYDRIN	026658424	173.25(A)(6) 173.25
1,2,5,6-TETRAHYDROCUMINIC ACID	056424874	
4,5,6,7-TETRAHYDRO-3,6-DIMETHYLBEN-ZOFURAN	000494906	
TETRAHYDROFURFURYL ACETATE	000637649	172.515
TETRAHYDROFURFURYL ALCOHOL	000097994	172.515

MAINTERM	CAS	REGNUM
TETRAHYDROFURFURYL BUTYRATE	002217336	172.515
TETRAHYDROFURFURYL CINNAMATE	065505251	
TETRAHYDROFURFURYL PROPIONATE	000637650	172.515
TETRAHYDRO-PSEUDO-IONONE	004433367	172.515
TETRAHYDROLINALOOL	000078693	172.515
TETRAHYDRO-4-METHYL-2-(2-METHYLPRO-PEN-1-YL)PYRAN	016409431	
5,6,7,8-TETRAHYDROQUINOXALINE	034413359	
TETRAMETHYL ETHYLCYCLOHEXENONE (MIXTURE OF ISOMERS)	977045694	172.515
1,5,5,9-TETRAMETHYL-13-OXATRICYCLO-(8.3.0.0(4,9))TRIDECANE	006790585	
2,3,5,6-TETRAMETHYLPYRAZINE	001124114	
THAUMATIN	053850343	
THEASPIRANE	036431728	
THEOBROMINE	000083670	
THIAMINE	000059438	PART 139 PART 137 136.115
THIAMINE HYDROCHLORIDE	000067038	182.5875 184.1875
THIAMINE MONONITRATE	000532434	182.5878 184.1878
THIAZOLE	000288471	
2-THIENYL DISULFIDE	006911519	
2-THIENYL MERCAPTAN	007774745	172.515
2,2'-(THIODIMETHYLENE)-DIFURAN	013678676	
THIODIPROPIONIC ACID	000111171	182.3109 181.24
THIOGERANIOL	039067806	
THIOUREA--PROHIBITED	000062566	189.190

MAINTERM	CAS	REGNUM
THISTLE, BLESSED (CNICUS BENEDICTUS L.)	977023134	172.510
THISTLE, BLESSED, EXTRACT (CNICUS BENEDICTUS L.)	977048228	172.510
THISTLE, BLESSED, EXTRACT SOLID (CNICUS BENEDICTUS L.)	977048240	172.510
THISTLE, BLESSED, OIL (CNICUS BENEDICTUS L.)	977048239	172.510
L-THREONINE	000072195	172.320
4-THUJANOL	000546792	
THYME OIL (THYMUS VULGARIS L. AND T. ZYGIS VAR. GRACILIS BOISS.)	008007463	182.20
THYME OLEORESIN	977029723	182.20
THYME (THYMUS SERPYLLUM L.)	977052371	182.10
THYME (THYMUS VULGARIS L.)	977052360	182.10
THYME, WILD OR CREEPING, EXTRACT(THYMUS SERPYLLUM L.)	977069967	182.20
THYMOL	000089838	172.515
TITANIUM DIOXIDE	013463677	73.575 175.105 175.210 175.300 175.380 175.390 176.170 177.1200 177.1210 177.1350 177.1400 177.1460 177.1650 177.2600 181.30
A&-TOCOPHEROL ACETATE	000058957	182.8892
A&-TOCOPHEROL ACID SUCCINATE	004345033	
TOCOPHEROLS	001406662	182.5892 182.3890 182.5890 182.8890 182.8892 184.1890
TOLUALDEHYDE GLYCERYL ACETAL (MIXED O-, M-, P-)	977041692	172.515

MAINTERM	CAS	REGNUM
TOLUALDEHYDES (MIXED O-, M-, P-)	001334787	172.515
TOLU, BALSAM, EXTRACT (MYROXYLON SPP.)	977075287	172.510
TOLU, BALSAM, GUM (MYROXYLON SPP.)	009000640	172.510
O-TOLUENETHIOL	000137064	
P-TOLYLACETALDEHYDE	000104096	172.515
O-TOLYL ACETATE	000533186	172.515
P-TOLYL ACETATE	000140396	172.515
4-(P-TOLYL)-2-BUTANONE	007774790	172.515
O-TOLYL ISOBUTYRATE	036438547	
P-TOLYL ISOBUTYRATE	000103935	172.515
P-TOLYL LAURATE	010024574	172.515
P-TOLYL 3-METHYLBUTYRATE	055066563	
P-TOLYL OCTANOATE	059558235	
P-TOLYL PHENYLACETATE	000101940	172.515
2-(P-TOLYL)-PROPIONALDEHYDE	000099729	172.515
O-TOLYL SALICYLATE	000617016	
TRAGACANTH, GUM (ASTRAGALUS SPP.)	009000651	133.133 133.162 133.178 184.1351 133.134 150.161 150.141 133.179
TRIBUTYL ACETYLCITRATE	000077907	172.515
TRICHLOROETHYLENE	000079016	173.290 172.560 175.105 177.1960
2-TRIDECANONE	000593088	
2-TRANS,4-CIS,7-CIS-TRIDECATRIENAL	013552960	

MAINTERM	CAS	REGNUM
2-TRIDECENAL	007774825	172.515
TRIDODECYL AMINE	000102874	173.280
TRIETHANOLAMINE	000102716	173.315(A)(3) 173.315 175.105 175.300 175.380 175.390 176.170 176.180 176.200 176.210 177.1210 177.1680 177.2260 177.2600 178.3120 178.3910
TRIETHYL CITRATE	000077930	182.1911 184.1191
TRIETHYLENETETRAMINE CROSS-LINKED WITH EPICHLOROHYDRIN	027754945	173.25(A)(6) 173.25
TRIFLUOROMETHANE SULFONIC ACID	001493136	173.395
2,4,5-TRIHYDROXYBUTYROPHENONE	001421632	172.190
TRIMETHYLAMINE	000075503	173.20
2,6,6-TRIMETHYL-1 AND 2-CYCLOHEXEN-1-CARBOXALDEHYDE	977045718	
P,A&,A&-TRIMETHYLBENZYL ALCOHOL	001197019	
4-(2,6,6-TRIMETHYLCYCLOHEXA-1,3-DI-ENYL)BUT-2-EN-4-ONE	023696857	
2,6,6-TRIMETHYLCYCLOHEXA-1,3-DIENYL METHANAL	000116267	
2,2,6-TRIMETHYLCYCLOHEXANONE	002408379	
2,6,6-TRIMETHYL-1-CYCLOHEXEN-1-ACE-TALDEHYDE	000472662	
2,6,6-TRIMETHYLCYCLOHEX-2-ENE-1,4--DIONE	001125219	
4-(2,6,6-TRIMETHYLCYCLOHEX-1-ENYL)-BUT-2-EN-4-ONE	035044689	
2,2,3-TRIMETHYLCYCLOPENT-3-EN-1-YL ACETALDEHYDE	004501580	

MAINTERM	CAS	REGNUM
3,5,5-TRIMETHYLHEXANAL	005435643	
3,5,5-TRIMETHYL-1-HEXANOL	003452979	
1,3,3-TRIMETHYL-2-NORBORNANYL ACETATE	013851111	
2,2,4-TRIMETHYL-1,3-OXACYCLOPENTANE	001193119	
2,4,5-TRIMETHYL-D&-3-OXAZOLINE	022694968	
2,6,10-TRIMETHYL-2,6,10-PENTADECAT-RIEN-14-ONE	000762298	
2,3,5-TRIMETHYLPYRAZINE	014667551	
2,4,5-TRIMETHYLTHIAZOLE	013623115	
2,2,6-TRIMETHYL-6-VINYLTETRAHYDROP-YRAN	007392190	
1,2,3-TRIS((1'-ETHOXY)ETHOXY)-PROP-ANE	067715826	
TRISODIUM NITRILOTRIACETATE	005064313	173.310
TRITHIOACETONE	000828262	
TRYPSIN FROM ANIMAL TISSUE	009002077	
L-TRYPTOPHAN	000073223	172.320
TUBEROSE, OIL (POLIANTHES TUBEROSA L.)	008024053	182.20
TUNU (CASTILLA FALLAX COOK)	977011543	172.615
TURMERIC (CURCUMA LONGA L.)	000458377	182.10 73.600
TURMERIC, EXTRACT (CURCUMA LONGA L.)	977083263	182.20
TURMERIC, OLEORESIN (CURCUMA LONGA L.)	129828291	73.615 182.20
TURPENTINE, GUM (PINUS SPP.)	009005907	172.510
TURPENTINE, RECTIFIED	888286199	172.510
TURPENTINE, STEAM DISTILLED (PINUS SPP.)	008006642	172.510

MAINTERM	CAS	REGNUM
L-TYROSINE	000060184	172.320
L-TYROSINE ETHYL ESTER HYDROCHLORIDE	004089070	
ULTRAMARINE BLUE	001302836	178.3970 73.50
2,4-UNDECADIENAL	013162464	
2,5-UNDECADIENAL	091254001	
2,3-UNDECADIONE	007493596	172.515
G&-UNDECALACTONE	000104676	172.515
UNDECANAL	000112447	172.515
UNDECANOIC ACID	000112378	
2-UNDECANOL	001653301	
2-UNDECANONE	000112129	172.515
10-UNDECENAL	000112458	172.515
2-UNDECENAL	002463776	
9-UNDECENAL	000143146	172.515
10-UNDECENOIC ACID	000112389	
UNDECEN-1-OL	000112436	172.515
2-UNDECEN-1-OL	037617031	
10-UNDECEN-1-YL ACETATE	000112196	172.515
UNDECYL ALCOHOL	000112425	172.515
N-UNDECYLBENZENESULFONIC ACID	050854949	173.315
UREA	000057136	184.1923
VALENCENE	004630073	
VALERALDEHYDE	000110623	172.515
VALERIAN ROOT, EXTRACT (VALERIANA OFFICINALIS L.)	008057496	172.510
VALERIAN ROOT, OIL (VALERIANA		

MAINTERM	CAS	REGNUM
OFFICINALIS L.)	008008886	172.510
VALERIC ACID	000109524	172.515
		173.315
G&-VALEROLACTONE	000108292	
DL-VALINE	000516063	
L-VALINE	000072184	172.320
VANILLA, EXTRACT (VANILLA SPP.)	008024064	182.20
		163.111
		163.112
		163.123
		163.113
		163.114
		163.117
		163.130
		163.135
		163.140
		163.145
		163.155
		163.150
		163.153
VANILLA, OLEORESIN (VANILLA SPP.)	008023787	182.20
		163.111
		163.112
		163.123
		169.175
		163.113
		163.114
		163.117
		163.130
		163.135
		163.140
		163.145
		163.155
		163.150
		163.153
		169.176
		169.177
		169.178
		169.180
		169.181
VANILLA (VANILLA SPP.)	977004060	182.10
		163.111
		163.112
		163.123
		169.175
		163.113
		163.114
		163.117
		163.130
		163.135
		163.140
		163.145
		163.155
		163.150
		163.153
		169.176
		169.177
		169.178
		169.180
		169.181

MAINTERM	CAS	REGNUM
VANILLIN	000121335	182.90
		163.112
		182.60
		135.110
		163.130
		163.123
		163.111
		163.113
		163.114
		163.117
		163.135
		163.140
		163.145
		163.155
		163.150
		163.153
VANILLIN ACETATE	000881685	172.515
VANILLIN ISOBUTYRATE	020665854	
VANILLYL ALCOHOL	000498000	
VANILLYLIDENE ACETONE	001080122	
VEGETABLE GUMS, OTHER THAN THOSE CFR LISTED	888286224	
VEGETABLE JUICE	977010584	73.260
VERATRALDEHYDE	000120149	172.515
VERBENOL	000473676	172.515
VERONICA (VERONICA OFFICINALIS L.)	977000831	172.510
VERVAIN, EUROPEAN (VERBENA OFFICINALIS L.)	977000411	172.510
VETIVER, OIL (VETIVERIA ZIZANIOIDES STAPF)	008016964	172.510
VETIVER (VETIVERIA ZIZANIODES STAPF)	977059703	172.510
VETIVERYL ACETATE	062563808	
VINYL ACETATE	000108054	172.892
O-VINYLANISOLE	000612157	
VINYL CHLORIDE-VINYLIDENE CHLORIDE COPOLYMER	009011067	172.210
P-VINYLPHENOL	002628173	
VIOLET LEAVES, ABSOLUTE (VIOLA ODORATA L.)	008024086	182.20

MAINTERM	CAS	REGNUM
VIOLET, SWISS (VIOLA CALCARATA L.)	977089103	172.510
VITAMIN A	000068268	PART 131 166.110 166.40 PART 133 182.5930 135.130 184.1930
VITAMIN A ACETATE	000127479	182.5933 184.1930
VITAMIN A PALMITATE	000079812	182.5936 184.1930
VITAMIN B-12	000068199	182.5945 184.1945
VITAMIN B COMPLEX AND SYRUP	977154087	
VITAMIN D	001406162	137.350 139.155 139.115 166.110 166.40 137.305 PART 131 137.235 137.260 184.1950
VITAMIN D-2	000050146	182.5950 184.1950
VITAMIN D-3	000067970	182.5953 184.1950
VITAMIN K	012001795	
WALNUT HULL, EXTRACT (JUGLANS SPP.)	977014382	172.510
WALNUT LEAVES, EXTRACT (JUGLANS SPP.)	977091987	172.510
WHEAT GLUTEN, VITAL	008002800	184.1322
WHEY	092129903	135.110 135.140 184.1979
WHEY, CONDENSED	977092015	184.1979
WHEY, DELACTOSED	092129936	184.1979
WHEY, DEMINERALIZED	977085883	184.1979

MAINTERM	CAS	REGNUM
WHEY, DRY	977032282	135.110 184.1979
WHEY, PARTIALLY DIMINERALIZED AND PARTIALLY DELACTOSED	977086364	184.1979A 184.1979B
WINTERGREEN, EXTRACT (GAULTHERIA PROCUMBENS L.)	977092742	
WINTERGREEN, OIL (GAULTHERIA PROCUMBENS L.)	068917759	
WOODRUFF, SWEET (ASPERULA ODORATA L.)	977070099	172.510
WORT	977042628	
XANTHAN GUM	011138662	133.162 172.695 133.178 133.134 133.133 133.124 133.179
2,5-XYLENOL	000095874	
2,6-XYLENOL	000576261	
3,4-XYLENOL	000095658	
XYLITOL	000087990	172.395
D-XYLOSE	000058866	
YARROW, HERB (ACHILLEA MILLEFOLIUM L.)	977000160	172.510
YEAST AUTOLYSATE	977046755	
YEAST, DRIED IRRADIATED	977052702	137.305
YEAST EXTRACT	008013012	
YEAST EXTRACT AUTOLYZED	977082782	
YEAST-MALT SPROUT EXTRACT	977011554	172.590
YEASTS	977030399	160.105 160.185 160.145
YEASTS, DRIED	977009361	172.896 139.122 139.155 137.235

MAINTERM	CAS	REGNUM
		139.115
YELLOW PRUSSIATE OF SODA	014434221	172.490
		573.1020
YERBA SANTA, FLUID EXTRACT (ERIODICTYON CALIFORNICUM (HOOK AND ARN) TORR)	977092731	172.510
YLANG YLANG, OIL (CANANGA ODORATA HOOK. F. AND THOMAS)	008006813	182.20
YUCCA, JOSHUA TREE (YUCCA BREVIFOLIA ENGELM.)	977083218	172.510
YUCCA, MOHAVE, EXTRACT (YUCCA SPP.)	977083207	172.510
ZEDOARY BARK, EXTRACT (CURCUMA ZEDOARIA (BERG.) ROSC.)	977039910	182.20
ZEDOARY (CURCUMA ZEDOARIA (BERG.) ROSC.)	977052575	182.10
ZEIN POWDER	009010666	184.1984
ZINC ACETATE	000557346	582.80
ZINC CARBONATE	003486359	582.80
ZINC CHLORIDE	007646857	582.80
		182.70
		182.5985
		182.8985
ZINC DITHIONITE	007779864	182.90
ZINC GLUCONATE	004468024	182.5988
		182.8988
ZINC METHIONINE SULFATE	056329421	172.399
ZINC OXIDE	001314132	182.5991
		582.80
		182.8991
ZINC STEARATE	000557051	182.5994
		182.8994
ZINC SULFATE	007446200	582.80
		182.5997
		182.90
		182.8997
ZINGERONE	000122485	172.515